S0-AOI-577

CONFUCIAN BIOETHICS

Philosophy and Medicine

VOLUME 61

ASIAN STUDIES IN BIOETHICS AND THE PHILOSOPHY OF MEDICINE 1

CONFUCIAN BIOETHICS

Edited by

RUIPING FAN

Center for Medical Ethics and Health Policy
Baylor College of Medicine
Houston, USA

KLUWER ACADEMIC PUBLISHERS
DORDRECHT / BOSTON / LONDON

A C.I.P Catalogue record for this book is available from the Library of Congress.

ISBN 0-7923-5723-X

Published by Kluwer Academic Publishers,
P.O. Box 17, 3300 AA Dordrecht, The Netherlands

Sold and distributed in North, Central and South America
by Kluwer Academic Publishers,
P.O. Box 358, Accord Station, Hingham, MA 02018-0358, U.S.A.

In all other countries, sold and distributed
by Kluwer Academic Publishers, Distribution Center,
P.O. Box 322, 3300 AH Dordrecht, The Netherlands

Printed on acid-free paper

Printed and bound in Great Britain by MPG Books Ltd., Bodmin, Cornwall.

TABLE OF CONTENTS

RUIPING FAN

INTRODUCTION: TOWARDS A CONFUCIAN BIOETHICS

I. "MUSEUM" BIOETHICS OR REAL BIOETHICS?

The title of this volume, *Confucian Bioethics*, may sound odd to some. It is odd to them not because they find Confucianism has lost its traditional strength in its homeland. It is odd because they doubt any essential relevance that Confucianism still has to contemporary society in general or to bioethical issues in particular. As the world changes, it seems that all traditional world views have been in retreat before a global cosmopolitan view. Confucianism is the tradition that seems to have declined most speedily in the 20th century. Even the so-called "last Confucian" has passed away (Alitto, 1985). For some, the only appropriate "ism" for the contemporary world is cosmopolitanism, because it attempts through reason alone to provide moral guidance to all people in all places. Confident in the creative power and justifying capacity of human reason, cosmopolitans hold that all particular traditional moral resources are rationally irrelevant to contemporary moral regulation. Indeed, cosmopolitanism intends to be independent from any particular tradition.

This volume illustrates an intellectual picture that differs from the view of cosmopolitanism. Instead of engaging in cosmopolitan bioethics, it offers particular Confucian perspectives concerning important bioethical issues. Specifically, this volume provides the Confucian views regarding the human body, health, virtue, suffering, suicide, euthanasia, "human drugs," human experimentation, and health care justice. These views are Confucian because they are derived from particular Confucian metaphysical, cosmological, and moral convictions and assumptions. They contrast with modern Western liberal perspectives in a number of important ways. If one is not sympathetic to basic Confucian metaphysical assumptions and/or moral convictions, one cannot find these views appealing. Indeed, the major authors for this volume are Chinese scholars studying or working in North American areas. They provide bioethical assumptions, arguments, and conclusions that are peculiarly Confucian, not cosmopolitan. Some, following the academic tendency of cosmopolitanism in the present time, would wonder why

Ruiping Fan (ed.), Confucian Bioethics, 1–23.

these authors cannot simply explore bioethical issues from a reasonable neutral stance, i.e., from a stance that is only based on reason and is not based on any particular tradition like Confucianism? Why is it necessary to pursue Confucian bioethics? And what is the epistemic status of Confucian bioethics?

These questions are not new. In fact, as early as the 1960s, historian Joseph Levenson already vividly described the triumph of cosmopolitan culture over the Confucian tradition. He predicted that the fate of Confucianism was inevitably decline:

> [t]he sageliness of Confucius may still be felt in China (or felt again), like Socrates' in Europe. But Confucian civilization would be as "historical" as Greek, and modern Chinese culture as cosmopolitan as any, like the western culture that reaches now, in paper-back catholicity, to "The Wisdom of Confucius." In a true world history, when all past achievements are in the museum without walls, everyone's past would be everyone else's; which implies that quite un-Confucian thing, the loss of the sense of tradition (1968, Part Three, p. 123).

The sense of tradition is lost in contemporary society because, from Levenson's historical perspective, no tradition can still stand as a whole. World civilizations are placed in "the museum without walls" only for visiting, not for living. When traditions are unavoidably fragmented in "the museum," cosmopolitanism arises in life.

The bits and pieces of valuable items left from each tradition may be incorporated into the melting pot of cosmopolitanism in the expectation of forming a unified moral system that can be justified by reason alone and be accepted by everyone alike. Accordingly, cosmopolitanism strictly speaking does not attempt to go without any traditional trappings; instead, it attempts to mix all and only rationally justifiable items from all traditions to shape a new comprehensive doctrine. The content of this doctrine is expected to cover only that which can pass the tribunal of reason. It is hoped that the "truth" from this doctrine is similarly presented, wherever people find themselves in the world, Beijing or New York, London or Singapore. Moreover, from this doctrine, people should have changed to "cosmopolitans" from traditional "communitarians." Consequently, to cosmopolitans, to do Confucian bioethics in the contemporary age is only to do "museum" bioethics. It may offer some useful bioethical bricks and tiles to the general building of cosmopolitan bioethics, but it cannot establish a real bioethics that people should follow as a particular way of life in preference to cosmopolitan bioethics. Apparently, any attempt to engage in Confucian bioethics as a real living moral system reflects only a nostalgic yearning for the good old days that will never return again. It seems that contemporary Confucian bioethics can only exist as "museum" bioethics.

This volume, however, offers a real Confucian bioethics. The authors of this volume are not interested in a "museum" Confucian bioethics because a "museum" bioethics is not real. It is not real because it is only to be visited, while a real bioethics is to be lived. For the authors of this volume, bioethics cannot find its significance except in a close tie to the real lives of individuals and their communities. The views they offer in the volume are a real bioethics because these views are lived by the people committed to the Confucian tradition. There are, after all, still Confucians in the contemporary world.

Confucian views are certainly not neutral views. The essays in this volume do not start from a neutral stance. Such a stance does not exist. Any moral view is from somewhere and is held by somebody. Attempting to derive a substantive stance solely from pure reason is illusory because pure reason is not substantive. Reason can play a constructive role in a moral system only through close combination with certain fundamental, content-full moral assumptions and premises, from which concrete bioethical views and visions are deduced, induced, or stimulated. Such fundamental, content-full assumptions and premises can only be found in particular moral traditions, such as Confucianism. However, it is impossible to unite basic assumptions from different traditions to make a coherent moral doctrine as cosmopolitans expect. These assumptions are mutually incommensurable.

It is necessary to pursue Confucian bioethics because its fundamental assumptions and premises are still in practice. Confucian values are still at home in East Asian areas such as Korea, Japan, Hong Kong, Taiwan, and mainland China. Although many people in these areas no longer describe themselves as Confucians, Confucian teachings continue to be deeply involved in their lives. Moreover, as people move and emigrate from their motherlands to other countries, non-geographically-isolated moral communities have formed in the contemporary world. Just as we can find Christian communities in East Asian areas, we can find Confucian communities in North American areas. Levenson was too pessimistic when he predicted that world traditions would exist only in the "museum." They are still existent in real lives. Instead of shrinking into one unified cosmopolitan culture, the contemporary world is witnessing the flourishing of different moral communities and real moral diversity. Confucianism is no longer restricted to East Asia. It is already present everywhere in the world

Understandably, problems can arise when different moral traditions confront each other regarding moral decisions. This is the case especially in the area of health care. We can find the most difficult and sometimes painful ethical conflicts between the East and the West in the issues of life, disease, suffering, and death. In North American areas, for example, such conflicts may result in confrontations between health care providers of the West and health care recipients of the East. A real Confucian bioethics can serve to clarify concepts, afford arguments, and give moral guidance. It is in this direction that the authors of this volume have focused their energies. Indeed, a real Confucian bioethics is practically inevitable.

II. AN EMOTIONAL RESPONSE?

Practical inevitability is not theoretical justifiability. The problem remains regarding the epistemic status of Confucian bioethics. Can traditional Confucian perspectives offer appropriate solutions to contemporary bioethical problems? Can Confucian bioethics be justified through rational philosophical argument? To the ears of many people from other traditions, Confucian bioethical views may sound parochial and even bizarre. Some might wonder, if bioethical exploration is primarily for the right answers to bioethical issues rather than for making practical surrender to available positions, why one should take traditional Confucian positions seriously. In fact, Levenson once depicted an embarrassing dilemma in which modern Chinese intellectuals had found themselves. On the one hand, according to Levenson, Chinese intellectuals are committed to the general. They want to seek the answers that are "true." On the other hand, however, they are also committed to the special. They want to hold the answers that are somehow "theirs." The first commitment brings them to intellectual alienation from Chinese tradition, while the second leaves them with an emotional tie to it (1968, p. xxxii). The reader of this volume must judge whether it is simply an emotional tie to the Confucian tradition that has brought the authors of this volume to attend to their particular explorations of bioethical issues.

It is pointless to argue whether the authors for this volume have an emotional tie to the Confucian tradition. Whether they have or do not have such a tie, emotional motivation is not philosophically interesting. What is philosophically important is to undertake rational exploration. Thus, the real matter here is not whether these authors have a commitment to special Confucian doctrines in addition to their commitment to general bioethical truth. Rather, it is whether these authors' commitment to bioethical truth should have rationally brought them to intellectual alienation from Confucianist understandings, as Levenson seemed to suggest. If the answer were "yes," then this volume would have been a nonstarter, reflecting nothing but an emotional response of the authors. However, in order to show that rationality requires these authors to stay away from the Confucian tradition, one must show that Confucianism is in conflict with rationality or bioethical truth. In other words, one must show that Confucianism is rationally to be rejected.

In fact, this has been the attitude of many modern Chinese scholars towards Confucianism since the May Fourth movement in 1919. "Down with Confucius and sons" has been sloganized in the Chinese intellectual circle. Many Chinese scholars, under the influence of modern Western social and political theories, asserted that Confucianism is primarily a feudal ideology in service of the totalitarian rulers of the past imperial dynasties. They held that Confucianism is the ultimate cultural factor responsible for the Chinese poverty and weakness in facing the powerful guns and cannons of the modern West. They concluded that Confucianism is fundamentally in opposition to De-Xian-Shen (Mr. Democracy) and Sai-Xian-Shen (Mr. Science), the two paradigm products of rationality. In short, it is their belief that Confucianism as a whole system must be abandoned in favor of new modern Western theories, even if it contains some particular good items that should be retained (see Tu, 1979, pp. 257-296).

Eighty years have passed since 1919. Both China and the rest of the world have witnessed tremendous alterations in their economy, politics, and morality. For one thing, after enormous frustrations and suffering experienced by the Chinese in this century, young Chinese scholars have again been able to study in Western countries since the early 1980s. Compared to their forerunners in the early century, these new Chinese sojourners in the West have fortunately had a more plentiful and peaceful opportunity to elaborate on modern Western theories and trace them back to ancient Western cultures. They have finally come to reflect on fundamental ethical issues while still keeping a distance from international political problems. In particular, perplexing bioethical challenges have given them a significant opportunity to reexamine traditional ways of life, including their own tradition. As several authors demonstrate in this volume, some dominant modern Western solutions to bioethical issues appear to be quite unreasonable. Such "modern" solutions have reminded them of the different teachings of traditional Confucianism. It is not that they wanted to find something peculiarly "theirs" that brought them back to Confucianism. It is their pursuit of bioethical truth that led them to Confucian perspectives.

These scholars never received systematic training in the Confucian classics as their forerunners did. They are not equal to their forerunners in comprehensive grasp of the large Confucian literature. Their advantage, however, lies in their more intimate experience of modern Western society and their closer examination of modern Western theories. The history of this century has taught them well that the most popular theory at a specific time may not be the true theory, while an abandoned doctrine may turn out to be the only defensible doctrine. It is in concrete ethical issues, especially in crucial bioethical issues, that profound metaphysical, cosmological, and moral disagreements show themselves most clearly. It is against this background that the authors of the volume find it rewarding to provide Confucian perspectives on some important bioethical issues. *Pace* Levenson, the essays in this volume are not a reflection of the authors' emotional tie to the Confucian tradition. Rather, they are rational responses based on the intellectual exploration of difficult bioethical issues.

III. CONFUCIANISM IS A PERSONAL HEALTH CARE SYSTEM

How does Confucianism approach personal health? Does exercising Confucian virtues, such as practicing humanity (*ren*), following traditional rites (*li*[b]), being filial (*xiao*) to one's parents, and obtaining knowledge (*zhi*[b]) of ancient classics, have anything to do with one's health? The common understanding of Confucianism would give negative answers to these questions. A Confucian is typically portrayed as weak and pale because he is busy studying, thinking, and teaching without spending time on physical exercise. In the first chapter of this volume, however, Peimin Ni argues that these are gross misunderstandings of Confucianism. From Ni's perspective, Confucians must exercise the moral virtues to obtain personal health because Confucianism holds that (1) personal health is a state of complete physical, mental, and social well-being and that (2) the virtues are the very qualities that define a healthy person. Thus, Ni contends, although Confucianism cannot be reduced to a personal health care system, it can be understood as a personal health care system. A person who does not perform the moral virtues is not a healthy person.

Ni's paper reminds the reader of the controversies between the negative and positive concepts of health in the discussion of the philosophy of medicine. Confucianism offers an understanding of positive health. Positive health is inescapably a matter of degree. A living person cannot be absolutely devoid of health in its positive sense. It is also difficult to be entirely healthy. From the Confucian view, a person's process of life ought to be a process of self-cultivation. Self-cultivation involves both internal (mental) and external (social) steps which can be understood as concrete methods for pursuing one's personal health. The internal steps include investigating things, extending knowledge, making the will sincere, and rectifying the heart-mind (*xin*[a]), while the external steps include regulating the family, governing the state, and making the entire world peaceful. By learning and exercising the virtues in carrying out these steps, one is pursuing and improving one's personal health. Indeed, from the Confucian perspective, a Daoist hermit is not quite healthy, even if he is physically well. He is not quite healthy because he stays away from the normal social relations (with the family, the state, and the world) that are necessary for his personal well-being. For Confucians, the well-being of a person is intrinsically connected to the well-being of the family, the state, and the entire world.

This Confucian holistic conception of personal health is well reflected in the traditional Chinese account of etiology. Chinese etiology, as Ni mentions, lists seven mental factors (delight, anger, sadness, pleasure, grief, fear, and fright) together with six natural factors (wind, cold, heat, humidity, dryness, and fire) as the most common causes of diseases. When these factors are excessive, insufficient, or unbalanced, health problems occur. According to Confucianism, it is by exercising the moral virtues that one can appropriately regulate these factors, especially the mental factors. For instance, Confucius observes that "those who are *ren* (humanity) are free from anxiety; those who are *zhi*[a] (wise) are free from perplexity; and those who are *yong* (courageous) are free from fear" (*Analects*, 14: 30). Moreover, if one is virtuous, one will follow the way of *zhong he* (equilibrium and harmony). The state of *zhong* is the state of the heart-mind that has "no stirring of pleasure, anger, sorrow, or joy," while the state of *he* is the state of the heart-mind that "has these feelings and they act in their due degree" (*Zhong Yong*, ch.1). Therefore, the Confucian virtues help maintain and improve personal health. This is why Confucius observes that "those who are *ren* have longevity" (*Analects*, 6: 21).

All this is profoundly manifested in the Confucian conception of the human body as a modality of *qi* (vital energy) and its internal moral sensibility, as Ellen Zhang illustrates in her essay. From the Neo-Confucian perspective, the human body is part of the onto-cosmological body that comprises the totality of Heaven, Earth, and myriad things, which are all made of *qi*. Indeed, for Confucians, the whole cosmos is a field of *qi*. There does not exist a personal god out there as the ultimate reality to create, preserve, and regulate the cosmos and human beings. Instead, for Neo-Confucians, *qi* itself can be considered the ultimate reality. However, this concept of *qi* cannot be understood by fitting into the dualism of Cartesian categories, either matter or mind. The language of particles, nuclei, electrons, hadrons, or even quarks is too materialistic to convey the sense of unspecifiable dynamism that *qi* signifies. The Leibnizian notion of monads, on the other hand, is too idealistic to express the Confucian meaning of this-worldly substance to which *qi* refers. Indeed, *qi* might be understood as both material and spiritual, existing always in a process of transformation and development. It is not being *per se*, but it is becoming that allows being to be more than merely something (Tu, 1992, p. 91). As a significant modality of *qi*, the human body is a microcosm of the whole cosmos to embody the evolution and transformation of Heaven, Earth, and myriad things in the cosmos. It is an open conduit through which the vital energy of Heaven and Earth flows, empowering the person fully to develop and realize oneself in participating in the great transformation of the cosmos.

Moreover, from the Confucian understanding, the sensibility and sensitivity of the human body is not only metaphysical and cosmological, but also moral. As Mencius states, everybody has flood-like *qi*, which is "exceedingly great and strong." It "is united with righteousness and the way." "By nourishing it with integrity and placing no obstacle in its path, it will fill the space between Heaven and Earth" (*Mencius*, 2A:12). Accordingly, for Confucians, to cultivate one's person is to nourish one's *qi*. To nourish one's *qi* is to exercise the moral virtues. And to exercise the moral virtues is to pursue personal health. Eventually, as Zhang summarizes, the goal of the Confucian self-cultivation is not Daoist immortality in a pure physical sense. Neither is it the Platonic immortality of the soul. It is to immerse one's body into the cosmos body by following the internal and external steps (investigating things, extending knowledge, making the will sincere, rectifying the heart-mind, regulating the family, governing the state, and making the entire world peaceful) through practicing the moral virtues, contributing one's vital energy to the *qi* field of the cosmos that profoundly affects others and, in turn, oneself. In this process one pursues and improves one's health. Accordingly, Confucianism is qualified as a health care system. The possession of the Confucian virtues is the possession of health.

This is not to say that Confucianism focuses only on the mental and social well-being of the person, while overlooking bodily health. Rather, Confucianism understands the inseparability of bodily health from psychological and social environments. It maintains a unity of physical flourishing, mental equilibrium, and societal peaceableness. Confucius observes, as quoted by Ni: "There are three things against which a gentleman (*jun zi*) is on his guard. In his youth, before his blood *qi* has settled down, he is on his guard against lust. Having reached his prime, when the blood *qi* has finally hardened, he is on his guard against strife. Having reached old age, when his blood *qi* is already decaying, he is on his guard against avarice" (*Analects*, 16:7).

IV. AN UPRIGHT DEATH IS NOT LAMENTABLE

Unlike the religious Daoist who struggles for physical immortality, Confucians understand that death is unavoidable. Because life and death are the normal transformation of flowing *qi* stream, death is as natural as life. According to Confucian metaphysics, human life comes into being when *qi* integrates into the human body, and death occurs when *qi* dissolves. As a modality of *qi*, the human body is constantly responsive to the vicissitudes of the ever-changing environmental *qi* field. And human life is inevitably susceptible to the evolutionary order of the entire cosmological body. Such evolutionary order is invested in everybody by Heaven and manifested by a series of the moral virtues, primarily *ren* (humanity). Accordingly, a normal human is one who follows the law of *qi* and nurtures the virtue of *ren*. "In life I follow and serve [Heaven and Earth]. In death I will be at peace" (Chang Tsai, 1963, p. 495). This is why Confucius comments that a man of *ren* does not have anxiety (*Analects*, 14:30). Indeed, a man of *ren* is not afraid of death.

Not only is a man of *ren* not afraid of death, he should actively embrace death when it is morally necessary. Ping-cheung Lo examines the Confucian views on suicide and their implications for euthanasia. Based on analyzing basic Confucian moral doctrines and exploring important historical cases, Lo formulates several Confucian theses and antitheses regarding death and suicide. First, from the Confucian view, one should give up one's life if necessary, either passively or actively, for the sake of upholding the cardinal moral values of *ren* (humanity) and *yi*[a] (righteousness). Preserving one's life is a good, but it is not the supreme good. Causing one's death is an evil, but it is not the supreme evil. The supreme good is a life lived in accordance with *ren* and *yi*[a], while the supreme evil is a life lived in violation of *ren* and *yi*[a]. Therefore, it is morally wrong to preserve one's life at the expense of *ren* and *yi*[a]. One should, rather, sacrifice one's life to uphold *ren* and *yi*[a]. Accordingly, Confucians do not generally object to committing suicide. In certain circumstances in which terminating one's life appears to be the only effective way to follow the requirement of *ren* and *yi*[a] (e.g., committing suicide to preserve a secret to save another's life, to avoid humiliation, or to protect one's dignity), Confucians hold that committing suicide is morally permissible, and sometimes is even morally admirable.

This Confucian view of suicide contrasts with the Western natural law tradition. Under the Western tradition one is never morally allowed to take one's life because it does not belong to oneself. Instead, as John Locke argues, men are God's property. They are God's workmanship "made to last during his, not one another's pleasure" (1980, p. 6). Therefore, "a man, not having the power of his own life, *cannot* ... take away his life, when he pleases" (1980, p. 23). In contrast, Confucians understand that men are the modalities of vital *qi* energy formed in the evolution and transformation of the *qi* field of Heaven and Earth in terms of the fundamental *qi* principle which manifests itself as the cardinal moral values of *ren* and *yi*[a]. In this process of transformation, human biological life or death by itself is not intrinsically good or evil. It is good or evil through its internal moral attachment to *ren* and *yi*[a]. Therefore, taking one's own life cannot be absolutely wrong in all circumstances. Committing suicide is morally meaningful when it is done in accordance with the principle of *ren* and *yi*[a].

This is not to say that Confucians do not appreciate the value of human life and generally support suicide. To the contrary, Confucianism always emphasizes the supreme value and nobility of humans among myriad things under Heaven, because only humans have been invested with the most excellent *qi*. Moreover, as Lo's Confucian antitheses illustrate, Confucians insist that one should broaden the scope of one's commitment. Instead of dying for a rather limited cause, one should live and die for an object of a higher order. Moreover, when there is no threat to one's life, and when the calling in life is clear, one should live on to fulfill one's vocation in spite of personal tragedy and undignified treatment. These ideas should have significant implications for the contemporary bioethical discussion of euthanasia. As Lo comments, on the one hand, Confucians may not absolutely object to voluntary, active euthanasia. When one has no more time and energy to continue one's project in life, when all palliative treatment does not work well, and when terminating one's life immediately is the only way of relieving one's suffering, it is hard to see how euthanasia can be opposed by *ren* in such a situation. On the other hand, however, a "good death" within the Confucian view on suicide is generally good for other-regarding reasons, not for self-regarding reasons as contemporary proponents of euthanasia understand the matter. Moreover, traditional cases of altruistic suicides may not offer support to those who request euthanasia for the sake of relieving the burden (emotional, financial, and otherwise) to others (family and society). This is because, from Lo's views, altruistic suicides in ancient China were usually intended to render a positive benefit to others. They were not encouraged in the form of negatively removing the "burden" to one's family or society. In short, Confucians do not object to euthanasia from any general principle. But they must be very careful about what reasons they have to perform active euthanasia in every particular case.

Lo's contrast between the self-regarding and other-regarding cases of suicide challenges the way the Chinese view of suicide has traditionally been understood in the West. As George Khushf in his responding essay points out, the Chinese view of suicide has normally been associated with a shame culture, and the act of suicide is viewed as self-regarding;

namely, as a way of saving face. Lo provides a valuable service by highlighting the deeper concerns with humanity and appropriateness that motivate the Confucian understanding of suicide. However, Khushf also argues that Lo's contrast between Confucian and Western views of suicide does not really convey the force of a profound classical Western liberal tradition. In particular, Khushf distinguishes two different strands of liberal thought. The first strand, termed an "anemic type of liberalism" by him, is one where individual liberty and limited government are derived from a principle of utility, based on individuals' preferences and self-determination. This principle is often an underlying ground for those who favor the legalization of assisted suicide and/or voluntary active euthanasia. The second strand of liberalism, Khushf contends, is the classical position (perhaps best represented by Locke) that works with the notion of inalienable rights and limited government. On this position, every individual has an inalienable right to life, liberty, and the pursuit of property. The source of this right is one with the source of one's humanity – human being and right are jointly rooted in a transcendent ground. Thus, this right places a limit upon the activity of an individual and a state. Since no individual can alienate life or liberty, and no state can claim arbitrary right over the life and liberty of its subjects, the prohibition against suicide is at the heart of this classical liberalism. For Khushf, Lo's Confucian antithesis on suicide ("one should live and die for an object of a higher order") should be modified and generalized. The "object of a higher order" should not be just emperor, dynasty, or even China. It should be Heaven itself. The result, from Khushf's understanding, is a Lockean inalienable right; namely, a right to life that cannot be alienated. In short, for Khushf, the Confucian view of suicide can be improved by the Lockean classical liberal notion of inalienable rights. On this view, neither suicide nor active euthanasia ought to be legalized.

The issue of active euthanasia is related to the issue of passive euthanasia, i.e., the issues regarding withholding or withdrawing medical treatment. This involves a judgment of medical futility. When should a therapeutic procedure be considered medically futile so that it should be withheld or withdrawn? Edwin Hui provides a Confucian ethic of medical futility. In the first place, he recognizes that no matter which approach one adopts to determine medical futility (be it physiological, qualitative, or statistical), such a determination cannot be value-free. A thirty percent reduction in tumor size can reasonably be considered as therapeutically effective or ineffective, depending upon one's worldview, belief system, and value judgment. From Hui's view, since Confucians see human persons in terms of the *qi* monism rather than a perspective such as the Cartesian mind-body dualism, they hold a high regard for the human body. Such a high regard, on the one hand, makes Confucians encounter considerable difficulty in accepting a futility determination for a patient who is in a so-called "permanent vegetative state." On the other hand, however, it also works in favor of not consenting to aggressive treatments, especially when they involve extensive surgical interventions which subject the body to disfigurement, dismemberment, or other forms of mutilation.

More importantly, Hui argues, Confucians hold a concept of social personhood which would characterize Confucian decision-making regarding medical futility. This Confucian concept of social personhood, from Hui's perspective, contrasts with the liberal humanist view of a person in the modern West. According to the liberal humanist view, a person is a self-determination entity. He is an autonomous individual decision-maker, master of his history, with full control of his destiny. The physician must disclose relevant medical information to the patient and obtain consent directly from the patient regarding medical procedures. The patient has a right to refuse any treatment, even if the treatment is considered useful by the physician and the family. The patient can also demand medically unestablished treatment and place extra pressures upon the physician. Such absolute patient self-sovereignty can even be asserted on the patient's behalf by another person if he is appointed by the patient or by a court as the patient's guardian. From Hui's view, such individualistic self-sovereignty places the issue of medical futility in a provocative confrontation of patient, physician, and family. It is an inappropriate approach to the complex issue of medical futility.

In contrast, the Confucian understanding of personhood emphasizes the social and relational nature of the human beings. The Confucian tradition never views a person as an isolated individual with absolute self-sovereignty. Instead, from the Confucian view, humans are distinguished from other animals exactly because humans have morally significant social organizations, such as family, community, and state. As a center of relations, a Confucian person must achieve harmonious cooperation with other human persons in these social institutions. This process of cooperation is the process of one's self-cultivation. On the one hand, the entire family must take care of a sick family member, including sustaining the burden of discussing treatment options with the physician and making ordinary therapeutic decisions for the patient. In clinical practice in relation to medical futility, Confucianism emphasizes mutual decision making of patient, family, and physician. No one has an absolute power to make a futility decision.

V. "HUMAN DRUGS" AND HUMAN EXPERIMENTATION

Jing-Bao Nie's essay offers a historical and ethical study of the traditional Chinese practice of "human drugs," i.e., materials from the human body. Although the majority of Chinese medicinals are of vegetable origins (herbs), animal and human drugs have also been used by Chinese physicians and been recorded in Chinese pharmacology since the second century. In particular, thirty-five human drugs are included in the great pharmacological work *Bencao Gangmu* (1597) by Li Shi-Zhen. This work is believed to reach the qualitative and quantitative climax in the development of Chinese materia medica. Among the fifty-two volumes of the *Bencao Gangmu*, Li devoted the entire last volume to human drugs. These include hair, pubes, fingernails, urine, blood, bones, placenta, gall, flesh, and so on. Traditional Chinese physicians believe that, for example, male pubic hair can treat snakebite, the husband's pubic hair can resolve the wife's difficult delivery, and the placenta can ameliorate impotence and infecundity.

Many people today would find these beliefs bizarre. How could the most brilliant Chinese physicians like Li Shi-Zhen believe that such strange drugs have medical therapeutic value? Nie reminds us of similar puzzles Thomas Kuhn faced when he read Aristotle's theories of natural phenomena, especially those concerning the motion of the objects, which are obviously nonsensical from the standpoint of modern physics. Chinese drugs similarly seem absurd because, Nie argues, we attempt to understand them from modern Western medical perspective. We are quick to take an attitude of modern scientism toward traditional doctrines, assuming that the modern scientific view is the only truth and that all other (traditional) views must be assessed against the "standard" scientific view. Nie contends that this attitude is inappropriate. Instead, he advocates an interpretive or hermeneutical approach to human drugs. First, this approach must go beyond the heavy shadow of scientism and recognize that traditional Chinese medicine holds a theoretical paradigm that is incommensurable from modern scientific discourse. Within the Chinese understanding of disease, health, *qi* monism, and yin yang interdependence and unity, the practice of human drugs and their therapeutic effects are coherently reasonable. Moreover, this interpretive approach tries to comprehend both commonalities and dissimilarities among different medical discourses, deciphering the cultural meanings of apparent absurdities or fundamental differences, such as those surrounding the human drugs. Finally, it must be sensitive to the historical and cultural context of traditional healing through the human drugs.

This is not to say that one must accept all the human drugs Li Shen-Zhen listed. To the contrary, as Nie discusses in detail, Li Shi-Zhen himself was very cautious and critical regarding human drugs. As a Confucian physician, Li understood medicine as an "art of *ren* (humanity)" and strongly insisted that the practice of the human drugs must be regulated by the Confucian moral principles of *ren* (humanity) and *yi*[a] (righteousness). Specifically, Li distinguished two types of human drugs. One type, such as hair, nails, urine, and urinary sediments, is morally permissible, while the other type, such as blood, gallbladder, bone, placenta, and flesh, is morally repugnant. Li might well believe in the medical efficacy of the latter type of the "drugs," but he entirely condemned their use based on Confucian moral point of view.

A fundamental reason underlying Li's moral objection to using these "drugs" is the Confucian understanding of the unique nobility of human beings that distinguishes them from any other materials. Given the nobility of human beings, for Li, it contradicts *ren* and harms *yi*[a] to use certain major components of the human body, such as bone and flesh, as therapeutic medicinals. In many cases Li's view is related to the Confucian understanding of the metaphysical nature of the human body and human relationships. For instance, regarding the fact that some people gathered human bones from abandoned graves to make medicine, Li argues that this behavior is morally wrong because (1) the bones of the dead still have sentience due to the qi responsiveness between the bone and the rest of the body, (2) even dogs do not eat the bones of dogs, and (3) it is an age-old custom and virtue to bury human bodies including bones. Moreover, Li is an enthusiastic spokesperson against using human flesh as medicine. Since the Tang dynasty (6th century, A.D.), cases had often been found in which children cut their flesh to be used as drugs to cure their parents. It was believed that the virtue of filial piety manifested in such actions together with the flesh played a powerful therapeutic role to their parents. Li explicitly condemned such practices. From his Confucian perspective, a child's body is from the parents and therefore is also the parents' body. Even if the parents suffer from a serious illness, Li asks, how can they take their own bones and flesh? For Li, such actions terribly violate the Confucian virtues of *ren*, *yi*[a] and filial piety.

Nie's interpretive construal of the traditional care of human drugs leads to an examination of our current practice of organ transplantation. As Nie observes, organ transplantation also involves the therapeutic use of parts of the human body. It can be seen as a modern form of the practice of human drugs. How do we make sense of such a form of practice? As Ronald Carson's commentary on Nie's essay points out, "the need to make moral sense of a practice arises only when the 'working sense' no longer suffices or when traditions of meaning clash" (p. 209). What is the moral difference between cutting one's flesh to cure one's mother and donating one's kidney to save one's mother, if they are both empirically effective? We still need a comprehensive Confucian theory of human drugs and organ transplantation.

Using human subjects in non-therapeutic medical experiments raises related ethical questions. Xunwu Chen explores a Confucian reflection on human experimentation. Given the Confucian understanding of the importance of knowledge to the public good, and given the risks to human subjects involved in non-therapeutic medical experiments, how should the Confucian treat such experiments? Chen begins with the Chinese story of the legendary farmer whom Confucians admire. The legendary farmer was believed to be the founder of traditional Chinese medicine. He tried various kinds of grasses and plants to learn the nature and function of them so that people could either avoid them or use them for medical purposes. It is said that he was poisoned seventy times by seventy kinds of grasses on one day. In modern terms, what the legendary farmer performed is a non-therapeutic medical self-experiment. From Chen's view, his action reflects the Confucian values. Confucians call for "investigating things" and "extending knowledge" (The Great Learning). These are necessary steps of self-cultivation. At the same time, they contribute to the public good. Moreover, according to Chen's view, the precondition of human experimentation is, for Confucians, non-violating of the principle of *yi*[a] (righteousness) and in line with *ren* (humanity).

For Chen, this requirement distinguishes Confucianism from utilitarianism with respect to human experimentation. While both

Confucianism and utilitarianism emphasize the possible promotion of the public good resulting from an experiment as a reason to justify the use of human subjects, the Confucian conception of the public good contains the moral requirement of *ren* and *yi*[a] that utilitarianism does not. A utilitarian justification of human experimentation would require that the social benefits to be gained from such experimentation outweigh the harms that will be caused to society, including harms to the human subjects. What is missing in such a justification is an intrinsic moral requirement of righteousness or justice that goes beyond the consideration of quantitative social benefit and harm. By contrast, the Confucian principle of *yi*[a] in line with *ren* demands that "one cannot impose upon others what one does not desire for oneself" (*Analects*, 12:2) and that "one ought to wish the good for others as the same that one wishes for oneself" (*Analects*, 6:30). This principle rules out any inhumane experiments, such as those that are not meant to serve but to destroy humanity, like making biochemical weapons. It also excludes experiments that use non-consenting human subjects. The Confucian principles require a defense of humanity. The fundamental goal for Confucians is always the realization of righteousness rather than maximization of benefit. Thus Confucians would demand that no human subject should be used in an experiment except through his/her own consent.

VI. THE CONFUCIAN PATTERN OF HEALTH CARE

Qingjie Wang explores the Confucian view of appropriate care for one's aged parents. Do adult children have a moral obligation to care for their elderly parents? Some Western philosophers, such as Norman Daniels and Jane English, argue that children do not have any more of such an obligation than any other person in the society. This is because, they claim, children do not ask to be brought into this world in the first place. In other words, they do not consent to forming the parent/child relationship. Thus, for these philosophers, there is a basic asymmetry between parental and filial obligations. The parental obligation of caring for their young children is a "self-imposed" duty, while the children's filial obligation of caring for their elderly parents is "non-self-imposed" and thus cannot be morally required. Consequently, according to this argument, traditional filial obligation is left either as children's volunteer responsibility or as a moral burden on the whole society.

Wang argues that this argument is misleading because it depends upon two suspicious presuppositions. First, it assumes that there is no consensual commitment from children to forming the parent/child relationship. Second, it assumes that filial obligation, if it can be justified as a moral obligation, must be based upon the voluntary consent of children. Wang argues that both of these assumptions are problematic. In the first place, although it is true that children do not give explicit consent to establishing the parent/child relationship, it is also true that they do not refuse to establish such a relationship. The fact is that a young child is not yet a moral agent able to give or refuse such consent. Accordingly, the real question here is not whether a child gives such consent, but rather whether a child *would* give such consent if he/she were a moral agent. For Wang, as long as one's life is worth living, it is more reasonable to assume that the child would give a consent than not. After all, one (i.e. the child) could always commit suicide to terminate one's life if one judges that it is not worth living. Moreover, Wang recognizes that we did give our consent to maintaining the parents/children relationship when we were young. We never wanted our parents to stop taking care of us. This situation, from Wang's view, still does commit children at least to some, if not full, filial obligation to their parents.

More importantly, Wang argues, even if children do not have explicit consensual commitments to the parent-child relationship, their filial obligation of taking care of their aged parents can still be justified in terms of the nature of moral obligation. From a contemporary Western understanding, a moral obligation must be grounded in the voluntary consent of relevant moral agents. This understanding currently underlies major accounts of the nature of moral obligation in the West. It is emphasized that as a free, rational, and autonomous moral person, one is responsible only for the consequences of those actions which one in the absence of coercion has intentionally committed. This account, Wang argues, confuses two different types of moral "ought" or "obligation." One type of moral obligation is generated solely by the intentional consent of competent moral agents. Wang terms this moral responsibility. The other type of moral obligation is determined by one's existential situation and the social roles that one plays. Wang terms this moral duty. Since we do not always choose our existential situations or social roles, our moral obligation cannot always depend on that to which we have consented. What is more important is that to which we ought to consent. For example, one did not choose to be a human being, one did not choose to be the brother of one's sister, and etc. But such existential and social characteristics commit one to certain moral obligations in the absence of one's consent. A human being is not only an autonomous being. He/she is first and foremost a historical, social, and communal being. For Confucians, it is superficial to account for basic moral obligations entirely according to voluntary consent, ignoring the metaphysical and existential conditions of human beings.

Wang offers the Confucian understanding of adult children's filial duty (*xiao*) to take care of their aged parents. Confucianism is notorious in its strong requirement of adult children respectfully to care for their aged parents, including looking after their everyday lives, satisfying their mental needs and making them happy, and sacrificing to their spirits after their passing away. Whether all parents and the elderly receive good care is taken by Confucians as substantial test of a good society and a good government. Filial obligation has also been regulated through the Chinese laws from ancient times to the present. It is not only a moral mistake but also a legal liability if one does not take care of one's elderly parents. Now how does the Confucian justify such a filial obligation?

The most important justification, from Wang's view, is the Confucian theory of the "rectification of names." For Confucians, names not only have the epistemological function of referring to something in reality. They are also operative. They carry within themselves normative requirements for actions. For instance, names such as "parent" and "child" not only indicate biological, familial, and social facts. They also possess the norms of being the parent and the son. They reveal specific familial and social relations as well as the inherent privileges and obligations of each party. Just as the norm of loving and taking care of their young children is coherently implied in the name of "parents," the norm of respecting and taking care of their aged parents is coherently embedded in the name of "children." If one does not take care of one's aged parents, one is not qualified to be called a son or a daughter. For Confucians, nothing seems more natural and fair than receiving care from one's parents when one is young and offering care to them when they are old. This is not because of "owing" or "paying debts" as some modern Western philosophers would hold. Rather, this is integral to the basic Confucian metaphysical conviction of human existence. For Confucians, human life is a continuous stream. One's parents can be seen as one's life in the past and one's children can be seen as one's life in the future. The Confucian filial obligation cannot be understood as an index of a causal relation between parents and children expressed in "owing" or "paying debts." Rather, it is a basic existential mode of human life.

Wang's argument suggests that it may not be easy to isolate a theory of health care justice from concrete conception of the good life. As I show in the last chapter of this volume, fashionable contemporary Western theories of social justice, egalitarianism, utilitarianism, redistributivism, and Rawlsianism, share an interesting characteristic: they all attempt to establish an account of justice for the structure of society in isolation from particular understandings of the good life. I identify this characteristic as "an intended separation." Especially, recognizing the intractable difficulty of the Enlightenment project of establishing a comprehensive system of morality to guide both individual and society, contemporary endeavors shift their focus to the basic structure of society. They attempt to disclose only an account of justice to regulate society, without addressing the divergent understandings of the good life held by different religions, cultures, and ideologies. They hope that even if one is not able through reason to discover a standard, content-full conception of the good life for all humans, one can still use rational argument to justify some substantive principles of justice to regulate the political constitution and economic arrangements of society. Under such "an intended separation," each of these theories attempts to justify its view of justice independently of any concrete premises from particular religions, metaphysics, ideologies, or conceptions of the good life. Each contends that its requirements of justice are compatible with all conflicting but reasonable accounts of the good life. Consequently, each argues that its views of justice ought to be accepted by all reasonable individuals and communities in contemporary pluralist societies. John Rawls' theory of justice and Daniels' application of it to health care distribution stand out as magisterial representatives of such endeavors.

My essay argues that a separation between rightness and goodness intended by these contemporary Western theories of justice is not feasible. Although such a separation represents a modest retreat from the comprehensive moral programs of the modern Enlightenment project, it cannot actually be carried out. For instance, Rawls begins with a "thin" theory of the good in order to construct his principles of justice. He holds that the primary goods such as liberty, opportunity, wealth and income are compatible with any reasonable, full conception of the good life. Daniels uses this theoretical approach (especially the requirement of "fair equality of opportunity" within Rawls' second principle) to health care.

He argues that since diseases impair an individual's normal function and therefore restrict the normal opportunity range otherwise available to the individual, they decrease the individual's fair share of the normal opportunity range. Accordingly, Daniels concludes, the state provision of health care is justified in order to maintain fairness for society's members by preventing and treating their diseases. This conclusion requires a particular sense of normality that Rawls and Daniels contend in order to argue that diseases diminish the normal opportunity range. A number of cases (such as those that may exempt one from military service) show that disease or disability may increase rather than decrease the range of one's opportunities. Rawls and Daniels must contend that these involve abnormal opportunities and thus do not count. Therefore, their so-called "thin" concept of opportunity presupposes a robust, content-rich distinction between "normal" and "abnormal." Because it is only in terms of particular understandings of the good life that once can find substantive standards for normality or abnormality, their concept of opportunity is compatible only with some, but not all, full conceptions of the good life. It is illusory for them to claim that their view of health care distribution is independent of any particular conception of the good life.

In contrast, Confucianism does not hold a "thin" theory of the good. Instead, it provides a cardinal principle of *ren* (humanity) to regulate the basic structure of society as well as to guide the individual toward a full-fledged good life. For Confucians, the pattern of health care distribution a society should adopt depends on what the best application of the principle of *ren* is. The principle of *ren* requires one to apply one's love to all humans, but only after taking into account distinctions, orders, and the relative importance of social roles. One should begin with one's family in the context of a local Confucian community. This is why adult children's filial obligation to their aged parents is morally so important that it must be appropriately discharged. Moreover, the Confucian classics emphasize that people in the local district should voluntarily sustain each other in sickness. It never suggests that the government may or should collect resources through heavy taxes to ensure equal and universal health care for everyone. To the contrary, Mencius clearly states that "a government of *ren* ... must make the taxes and levies light" (*Mencius*, 1A: 5: 3). People should be left in their local community, with their own resources, freely and cooperatively to pursue an appropriate pattern of health care for themselves.

VII. CONCLUDING REMARKS

The unfeasibility of the "intended separation" between the right and good held by contemporary Western theories of justice is heuristic. It lends theoretical credence to Confucianism. The Confucian account of justice is not uniquely parochial because of its close connection with a particular understanding of the good life. The truth of the matter is that the so-called neutrally applicable theories, such as Rawls' theory, are not neutral at all. They presuppose specific understandings of the good life, although in a more hidden and even fragmented way. This certainly should not establish them as more reasonable or as having more universal claims.

No one denies the sociological fact of ethical diversity and plurality in the contemporary world. But cosmopolitans want to offer a global ethics justifiable by reason and acceptable to everyone. They want, at least, to establish a theory of justice as universally applicable, even if they have to leave full conceptions of the good life unaddressed. The former is a strong version of cosmopolitanism, while the latter a weak version. This volume suggests that neither is workable. Confucian bioethics stands out as a significant communitarian bioethics that offers a coherent way of engaging the good life that is as rationally justifiable as any other particular bioethics. The components of the good and the right in its system are consistently linked to each other in such a way that they cannot be isolated from each other, something not achieved by any other bioethical theory. Accordingly, instead of the fashionableness of cosmopolitanism, Confucian communitarianism offers a more coherent and profound experience of the moral life (Tu, 1989).

Center for Medical Ethics and Health Policy
Baylor College of Medicine
Houston, Texas

REFERENCES

Alitto, G. S.: 1986, *The Last Confucian: Liang Shu-ming and the Chinese Dilemma of Modernity*, University of California Press, Berkeley.

Chan, W. (ed.): 1963, *A Source Book in Chinese Philosophy*, Princeton University Press, Princeton.

Chang Tsai: 1963, 'The western inscription,' in W. Chan (ed.), *A Source Book in Chinese Philosophy*, Princeton University Press, Princeton, pp. 494-495.

Confucius: 1979, *The Analects*, D. C. Lau (trans.), Penguin Books, New York.

Levenson, J. R.: 1968, *Confucian China and Its Modern Fate: A Trilogy*, University of California Press, Berkeley.

Locke, J.: 1980, *Second Treatise of Government*, C. B. Macpherson (ed.), Hackett Publishing Company, Inc., Indianapolis.

Mencius: 1970, *Mencius*, D. C. Lau (trans.), Penguin Books, New York.

The Doctrine of the Mean, 1963, in *A Source Book in Chinese Philosophy*, Wing-Tsit Chan (trans. & comp.), Pinceton University Press, Princeton.

The Great Learning, 1963, in *A Source Book in Chinese Philosophy*, Wing-Tsit Chan (trans. & comp.), Pinceton University Press, Princeton, pp. 85-94.

Tu, W.: 1979, *Humanity and Self-Cultivation: Essays in Confucian Thought*, Asian Humanities Press, Berkeley.

Tu, W.: 1989, *Centrality and Commonality: An Essay on Confucian Religiousness*, a revised and enlarged edition, State University of New York Press, Albany.

Tu, W.: 1992, 'A Confucian perspective on embodiment,' in *The Body in Medical Thought and Practice*, Drew Leder (eds.), Kluwer Academic Publishers, Dordrecht, pp. 87-100.

PART ONE

BODY, HEALTH AND VIRTUE

PEIMIN NI

CONFUCIAN VIRTUES AND PERSONAL HEALTH

If we were to conduct a survey on the characteristics of the typical Confucian, few people would list "healthy" as one of these characteristics, along with moral virtues such as benevolence, following traditional rites, being filial to parents, having knowledge of ancient classics, etc. In common understanding, Confucianism has little direct connection with personal health. The virtues listed above are all considered to be about one's mental and ethical qualities, important in social and political realms, but irrelevant to personal health. Contrary to an image of a healthy person, a Confucian is typically portrayed as weak and pale because Confucians are always busy studying, thinking, and teaching, while spending little time on physical exercises. Some facts may have contributed to this misconception. For instance, in the *Analects*, Confucius' disciple Zi Lu was addressed by an old farmer as "you who with four limbs do not toil" (*Analects*, 18:7). The editors of a book titled *A Grand Compilation of Practical Chinese Health Preservation Theories and Prescriptions* were able to find only two passages from the entire *Analects* that are applicable to health care according to their understanding of "health" (Shi Qi, Lu Mingfang, 1990, p. 85) – one on eating (*Analects*, 10:8), and the other on guarding against one's desires (*Analects*, 16:7). They would probably regret that even these two are not purely on health care – one is tainted by some descriptions of Confucius' aristocracy, and the other with a color of moral preachment, which is normally considered extrinsic to health care.

This, however, is a gross misunderstanding of Confucianism. I shall show in this paper that to Confucians, the moral virtues they advocate are means to obtain personal health. If we take the concept of health in the positive sense, namely, not merely as an absence of disease but as a state of more complete well-being,[1] we can understand the whole of Confucianism as, though not reduced to, a system of health care. Furthermore, Confucian moral virtues can be understood as qualities that define a healthy person.

Ruiping Fan (ed.), Confucian Bioethics, 27–44.

I. CULTIVATION OF THE PERSON

One of the four major Confucian classics, the *Great Learning* (*Da Xue*), summarizes the whole Confucian program into the following eight interdependent steps, with each subsequent step contingent upon and as necessary consequence of its precedent: investigate things, extend knowledge, make the will sincere, rectify the heart-mind, cultivate the person, regulate the family, govern the state well, and bring peace to the world (*ge wu, zhi zhi, cheng yi, zheng xin, xiu shen, qi jia, zhi guo, ping tian xia*). Within this program, the step concerning the cultivation of the person (*xiu shen*) is a transitional link between what is more internal or personal and what is more external or social. Starting from a proper understanding of the term "*xiu shen*" and its related concepts, we shall be able to see how the whole Confucian project is intrinsically one of health care.

The Chinese word "*shen*[a]," typically translated as "person," is close to, but not equivalent to the English word "body." Sometimes we find Chinese expressions that put *shen*[a] and *xin*[a] (heart-mind) in contrast, as in the expressions "*shen xin jiao cui*" (both the heart-mind and the body are exhausted) and "*shen bu you ji*" (the body is out of one's own control), though other times the word *shen*[a] means more than the body, as in "*yi shen shi fa*" (risk one's own person to test the power of law). This fact indicates that the Confucian "*xiu shen*" does not at all exclude the cultivation or care of the body. As Tu Wei-Ming says, to the Confucians the body is "not a servant, a means, a transition, or a shell; it is the embodiment of the person" (Tu, 1984). As an embodiment, *shen*[a] is primarily the bodily aspect of the entire person. It can represent the entire person because it is the expression of the overall state of the person. In that sense, the mind or the mental aspect of the person is not excluded from *shen*[a]; it is embodied in it.

On the other hand, *xin*[a], the rough equivalent of the English word "mind," etymologically means "heart," and it still serves the function of both "mind" and "heart" in the Chinese language. That means that for Confucians *xin*[a] is not entirely incorporeal. Unlike the Cartesian mind, which is an ontological entity distinct from the body, the Confucian *xin*[a] is a bodily organ that has the function of thinking and feeling. The word "*xin*[a]" is used, when in contrast to the word "*shen*[a]," in the sense of the mental aspect (not a mental entity) of the person.

From this brief analysis of the meanings of "*shen*[a]" and "*xin*[a]," we find that *xiu shen*, the cultivation of the person, refers to the cultivation of the entire person as it is displayed in the embodiment of the person. Yet it does not exclude the cultivation of the heart-mind. When we look at *xiu shen* in connection with its preceding steps, namely *zheng xin* (rectify the heart-mind) and *cheng yi* (make the will sincere), we see more clearly that, according to Confucianism, the cultivation of the heart-mind is not at all separate from the cultivation of the bodily aspect of the person. To the contrary, it is an essential part of this project. Since the heart-mind is "the governing part of the person" (Zhu Xi, *Zhu Zi Yu Lei*, vol. 12), it is a key to one's physical well-being. Following Mencius (*Mencius*, 6A:11), Zhu Xi claims that

> The key to learning is to search for the peace of the heart-mind. [Here the word 'learning' is also to be taken broadly as not merely an intellectual learning of 'what;' it is also a practical learning of 'how.'] If the heart-mind is flowing everywhere without restraint, what can you rely on for governing [the person]? Other efforts are all slow and not effective. You must first hold your own heart-mind still, be determined not to let it be massed up, you will then naturally have brightness everywhere and will not fall short of resource (Zhu Xi, *Song Yuan Xue An – Hui Weng Xue An Shang*).

The *Great Learning* also explains:

> When the heart-mind is absent, one looks but does not see, listens but does not hear, eats but does not know the taste. That is why it is said that to cultivate the person, one must rectify the heart-mind (ch.7).

Today, ordinary health care programs are deeply grounded in the Cartesian mind-body dualism. They treat a human being as little more than a body, and the body as little more than a lump of different parts. On the other hand they take the mental cultivation of a person to be extrinsic to health care, as something that helps health care practice from without, rather than something that constitutes health care practice itself from within. Unlike those programs, the Confucian program envisions a healthy person in complete state of well-being. As a holistic project, Confucian *xiu shen* aims not merely at a cultivation of the mind, the intellectual and moral faculty of the person, or merely at a cultivation of the body as an aggregate of physical organs and limbs. It involves all of them. This holistic conception of health care does not mean that

Confucians are simply ambitious; rather, Confucians understand the person as a whole individual, and they conceptualize one's physiological, psychological, moral, and even social states as all closely related with one another and mutually affected by one another. For Confucians, it is impossible for a person to have an agitated heart-mind, and yet still be healthy. This follows from two reasons. First, the heart-mind is an inseparable part of the person, so the concept of being healthy should include mental health as well. Second, the heart-mind is the dominant part of the person; therefore, the state of the heart-mind inevitably affects the rest of the body. When the heart-mind is in an unhealthy condition, the body cannot be well for long; and when the body is in poor condition, proper care of the heart-mind can always help the recovery of the body. This is why the *Great Learning* states: "Only when the heart-mind is rectified, the person (*shen*[a]) is cultivated" (ch.1).

This kind of holistic conception of health care is typical to Chinese thought in general, and it is well reflected in traditional Chinese medical theory and practice. Traditional Chinese doctors are trained in such a way that they do not look at the parts of a person as isolated. They understand that the organs in our body mutually affect each other, and that they are all affected by our mental-psychological states. One etiological theory of traditional Chinese medicine lists delight, anger, sadness, pleasure, grief, fear, and fright (*xi, nu, ai, le, bei, kong, jing*) as seven internal causes of health problems, along with six external causes – wind, cold, heat, humidity, dryness, and fire (*feng, han, shu, shi, zao, huo*). Another popular traditional Chinese medical theory tells us that excessive anger hurts the liver, excessive pleasure hurts the heart, excessive thinking hurts the spleen, excessive sadness hurts the lungs, and excessive fear hurts the kidneys. A good doctor, therefore, often goes beyond the patient's physiological condition and cares about the patient's mental-psychological condition as well. Since the heart-mind is the governing part of the person, it is also the key to many health problems. Rectifying the mind can prevent such problems from occurring, rather than being bandage-like and waiting for the problems to occur before fixing them. Moreover, it also solves existing problems by removing their root, rather than merely trying to ease the symptoms.

In this regard, Confucianism is a typical Chinese health care system as it is particularly strong in cultivating the heart-mind. In addition, it goes beyond the personal level, and sees the connection between the well-being of an individual and the well-being of the family, the state, and the whole world (*tian xia*). The Western Han Dynasty Chinese classic on medicine *Huang Di Nei Jing* states that "the sages do not wait until the sickness is there to cure the sickness, they cure it before it takes place ... If one waits until the sickness is there and then uses medicine to cure it, that is no different from waiting until one is thirsty and then starting to dig a well" (*Su Wen – Si Qi Tiao Shen Da Lun*). The Tang Dynasty Confucian-doctor Sun Si Miao makes another striking statement: "A superior doctor takes care of the state, a mediocre doctor takes care of the person, an inferior doctor takes care of the disease" (*Bei Ji Qian Jin Yao Fang*, vol. 1).

II. *REN* AND *ZHONG YONG*

Of course, the above quote does not mean that the supreme doctor does not take care of the disease or the person. It means that the supreme doctor is not limited to taking care of the disease or the person. In order to remove the disease, one has to treat the person as a whole, and in order to secure the well-being of the person, the social environment must be made suitable. As we have seen from the eight step program outlined in the *Great Learning*, a good social environment comes from well cultivated individuals, the way to cultivate the person is to cultivate the heart-mind, and the way to cultivate the heart-mind is to make one's thoughts and intentions sincere (*chen yi*). To Confucians, this means primarily to make the person virtuous, from the inside.

Of all the Confucian virtues, *ren* (often translated as human heartedness, humanity, benevolence, etc.) is the most central. Confucius, in a statement that clearly indicates the relationship between *ren* and personal health, states that "those who are *ren* have longevity" (*Analects*, 6:21). "The reason that those who are *ren* often have longevity is," Han Dynasty Confucian Dong Zhong Shu explains, "that they are not greedy of external things and they are tranquil and pure internally; their heart-mind is peaceful, harmonious, and is not out of balance; they nourish their person with the best things from Heaven and Earth" (*Chun Qiu Fan Lu – Xun Tian Zi Dao*). This explanation naturally refers us to the famous instruction of Confucius: A person of *ren* "overcomes the self" (*ke ji*) (*Analects*, 12:1). It tells us clearly that this "overcoming the self" should not be taken simply as preaching for a morality of altruism. It is a

Confucian method of health care. A passage in the *Analects* shows the linkage between "overcoming the self" and health care very clearly:

> There are three things against which a gentleman (*jun zi*) is on his guard. In his youth, before his blood *qi* has settled down, he is on his guard against lust. Having reached his prime, when the blood *qi* has finally hardened, he is on his guard against strife. Having reached old age, when the blood *qi* is already decaying, he is on his guard against avarice (16:7).

According to *Huang Di Nei Jing*, "Blood *qi* is the *shen*[b] (spirit) of a human being" (*Su Wen – Ba Zheng Shen Ming Lun Pian*). Without *shen*[b], one cannot live. Guarding against one's own lust, strife, and avarice therefore constitutes advise for protecting one's own spirit and one's essential well-being, and is not merely moral preaching.

Along the same line, Mencius says, "In caring for the heart-mind, nothing is more effective than having fewer desires" (*Mencius*, B:35). Qing Dynasty scholar Jiao Xun quotes Chuang Tzu to support and explain this saying:

> In the state of Lu there was a man named Shan Bao, who lived in rocky mountain and drank plain water, and never competed with others for material gains. His skin looked still like a young child when he was seventy. Unfortunately he was attacked and eaten by a hungry tiger. There was another man named Zhang Yi, whose footprints were at the doors of all the rich and the famous. He died in a disease of internal fever. Bao cared for the internal and the tiger ate him from without; Yi cared the external and the disease hit him from within (Jiao Xun, *Meng Zi Zheng Yi*, vol. 2, p. 1017).[2]

Even though Shan Bao was eaten by a tiger, he was very healthy at seventy, and that is attributed to his having little desire for material satisfaction. *Huang Di Nei Jing* also attributes short living to indulgence in desires:

> I heard that people in ancient times often lived over one hundred years, and at that age they were still able to move around without much difficulty. People today start to have trouble moving around even at the age of fifty. Is that simply because the times are different? Does it mean that humans are about to extinguish? Qi Bo says, those ancient people who understood the *Dao* followed the movement of *yin* and *yang*, harmonized methods and limits, controlled their eating and

> drinking, had regularity in getting up and sleeping, never exhausted themselves unnecessarily. That is why they were able to keep fit both in shape and in vitality, lived up to the natural limit of human life, and died after one hundred years old. People today are different. They use alcohol as regular drink, take extremity as constant; they go to bed after they got drunk, and exhaust their energy in love-making; they use up their precious vitality, and have no idea of holding up and accumulation. They do not guard their spirit from time to time, and they want immediate pleasure for satisfaction. They live opposite to the principles of life and happiness, and have no regularity in getting up and sleeping. That is why they are declined at the age of fifty" (*Su Wen – Shang Gu Tian Zhen Lun*).

Zhu Xi even believes that "the tens of thousands of teachings of the [Confucian] sages are nothing but instructing people to *be aware of the principles of Heaven and eliminate human desires*" (*Zhu Zi Yu Lei*, vol. 12). This famous slogan is often considered a social and political principle, but actually it is primarily a way of cultivating personal well-being. *The Doctrine of the Mean* states, "the superior man is cautious over what he does not see and apprehensive over what he does not hear" (ch.1). Wang Yang Ming explains: "If one can truly be 'cautious over not-seen and apprehensive over not-heard,' and be always like that, one will then be able to retain one's spirit, one's *qi*, and one's vitality, and the Daoist teachings of the so called immortality and long-lasting vision is also already inside." That is why "roughly speaking, care for moral virtue and care for the person are one and the same thing" (*Wang Wen Cheng Gong Quan Shu*, vol. 5).

Confucianism never advocated total elimination of all human desires. By "eliminating human desires," Zhu Xi means the elimination of desires of material possession, or the sickness of the heart-mind.[3] "As for the desire to eat when hungry, the desire to drink when thirsty, how can we live without those desires!" (*Zhu Zi Yu Lei*, vol. 94). Zhu Xi knows very well the importance of Confucius' teachings on *zhong yong,* the doctrine of the mean. He is actually the person who selected *The Doctrine of the Mean* from *Li Ji* (*The Book of Rites*), to be one of the four major Confucian Classics. "*Zhong*[a]" means equilibrium, centrality, "having no one-sidedness, not excessive nor insufficient," and "*yong*" means "commonality" (Zhu Xi, *Zhong Yong Zhang Ju Zhu*. See Tu Wei-Ming 1989, 16-7). *The Doctrine of the Mean* says: "When equilibrium and harmony are realized to the highest degree, Heaven and Earth will attain their proper order and all things will flourish" (ch.1). Here "all things" certainly include human beings. *Zhong yong* as a virtue is therefore extremely important to health care. Dong Zhong Shu says, "There is no virtue greater than harmony, there is no path more correct than equilibrium. Equilibrium is the beauty of Heaven and Earth, and is a sensible principle that the sages retain ... One who cares for his own person with equilibrium and harmony will reach the natural limit of the human life span" (*Chun Qiu Fan Lu – Xun Tian Zhi Dao*). Confucius himself also says, "as a virtue, nothing is greater than equilibrium and commonality" (*Analects*, 6:29).

The principle of equilibrium is applied extensively in Chinese medicine, such as the balance of *yin* and *yang* in the body. *Huang Di Nei Jing* says, "those who have the balance of *yin* and *yang* are called balanced persons ... balanced persons will not be sick" (*Su Wen – Tiao Jing, Ling Shu – Zhong Shi*). This is an excellent indication that the Confucian virtue of *zhong yong* is a healthy state for human beings. Today, the Chinese still use the concepts of *yin* and *yang*, and search for the balance of the two by using herbs and choosing special diets. In addition, Chinese doctors still utilize these concepts in their diagnoses. The statement, "when *yang* is weak, one is outwardly cold; when *yin* is weak, one is internally hot" (*Huang Di Nei Jing – Su Wen – Tiao Jing*), remains one of the basic principle guidelines of Chinese medical practice. Typically the conception of *yin yang* balance is applied so broadly that most health problems can be explained as the imbalance of the two forces or elements.

III. *LI*, *CHENG*, AND *QI*

The measure of equilibrium and commonality is extremely difficult to grasp. Confucius believes that everything is changing, and different situations affect where equilibrium and commonality are to be found. One way to cope with the difficulty is to follow *li*, traditional rites or proprieties. "To return to the observance of the rites" is also one of the ways that Confucius uses as instruction on how to be *ren*. The master says to "restrain from looking at, listening to, saying, and doing anything that is not in accordance with rites" (*Analects*, 12:1). He believes that the rites are repositories of ancient wisdom and insights into what is appropriate and conducive to one's health and overall well-being. China has developed an enormous array of rites that range over all aspects of human life. These rites can help one to know how a properly restrained person should live and behave. Xun Zi considers the rites to be so useful that, he says, "in cultivating one's *qi* and cares for the heart-mind, no path is better than following the rites." According to Xun Zi:

> In using blood *qi*, will and intention, knowledge and deliberation, one always gets through smoothly if one follows the rites, and gets frustration, confusion, or detainment if one does not. In eating and drinking, clothing, residing, moving or resting, one always gets harmony and comfort if one follows the rites, and gets stumbled or sickness if one does not. In facial expression, attitude, going and leaving, approaching and appearing, one is graceful if one follows the rites, and one is eccentric and crude if not. Therefore, humans cannot live without rites, things cannot be accomplished without rites, and states cannot have peace without rites" (*Xun Zi*, ch.2 – "*Xiu Shen*[a]").

The importance and the power of *li* for bringing health has been forgotten or misunderstood by many. To some, *li* is a bondage that limits human freedom; to others, *li* is a set of social norms one has to obey to be a civilized person; to still others, *li* is just a symbol of tradition, convention, or social habit. Indeed, few people today would understand *li* as Xun Zi did. There is something in this passage that is quite foreign to today's "scientific mind," yet very crucial for understanding Confucianism. There is a "magical dimension," as Fingarette puts it (Fingarette, 1972, p. 5):

> I see you on the street; I smile, walk toward you, put out my hand to shake yours. And behold –- without any command, stratagem, force ... you spontaneously turn toward me, return my smile ... (Fingarette, 1972, p. 9).

Examples like this, Fingarette points out, show a magical power of *li* –- "the power of a specific person to accomplish his will directly and effortlessly," without physical coercion (Fingarette, 1972, p. 8).

The "magical power" of *li*, in Fingarette's reading, exists in the interpersonal dimension. He has noticed how, according to Confucius, when one performs *li* in a sincere way, one can affect others.[4] But it exists in a personal dimension as well. As a person is able to regulate oneself so that one sincerely does not look at, listen to, say, and do anything that is not in accordance with rites, one is in a well balanced state of heart-mind, and that leads to, or rather is itself, a healthy state of the person.

The traditional rites give one very specific instructions on how to behave. They are therefore extremely helpful for those who do not intuitively know how to care for their own well-being. But the point is not simply to perform the outward procedures. Without the agent's sincere presence, an outward form of a body gesture or a ceremonial procedure can hardly be better than meaningless performances; the rites are proper ways in which sincerity can be presented. Consider, for example, a hand-shake. The effect of a hand-shake differs according to whether it is done with sincerity, with reluctance, or entirely without one's own willingness (as in the case of one's hand being forced to touch another's). *The Doctrine of the Mean* puts sincerity (*cheng*) on a very high place: "Sincerity is the way of Heaven, The attainment of sincerity is the way of men" (ch. 20, trans. Legge, p. 413).

Sincerity means oneness in one's mind. "It is said in the *Book of Poetry,* 'Although the fish sink and lie at the bottom, it is still quite clearly seen.' Therefore the superior man examines his heart, that there may be nothing wrong there, and that he may have no cause for dissatisfaction with himself" (*The Doctrine of the Mean*, trans. Legge, pp. 431-2).

> The power of sincerity is tremendous: It is only he who is possessed of the most complete sincerity that can exist under Heaven, who can give its full development to his nature. Able to give its full development to his own nature, he can do the same to the nature of other men. Able to give its full development to the nature of other men, he can give their full development to the nature of animals and things. Able to give their full development to the natures of creatures and things, he can assist the transforming and nourishing powers of Heaven and Earth. Able to assist the transforming and nourishing powers of Heaven and Earth, he may with Heaven and Earth form a ternion" (*The Doctrine of the Mean*, trans. Legge, p. 416).

Or, as the Song Dynasty Confucian Tang Zhong You says: "Retain one's nature like caring for an infant, cultivate the heart-mind like cultivating a young plant, driving one's *qi* like driving a horse, guarding against desires like guarding against flood, treating material gains like treating invaders. All these can be summed up in one word: Sincerity" (*Song Yuan Xue An – Shuo Zhai Xue An*). This sincere presence is not something one generates at the time of performance. Without a long cultivation, there is no way one can reach this kind of oneness and clarity in one's mind. This is why *The Doctrine of the Mean* says, "that wherein the superior man cannot be equalled is simply in what the others cannot see" (ch. 33, my trans.).

Here Mencius' theory of caring for *qi* becomes quite relevant. *Qi* means, roughly speaking, "vital force" or "energy."[5] "*Qi* is what pervades and animates the body" (*qi, ti zhi chong ye*), and "the will (*zhi*[a]) is the leader of the *qi*" (*zhi, qi zhi shuai ye*). "When *zhi*[a] is there, *qi* comes" (*Mencius*, 2A:2). *Qi* cannot be obtained by instant decision making. "It is produced by the accumulation of righteousness and cannot be appropriated by anyone through a sporadic show of rightness. Whenever one acts in a way that falls below the standard set in one's heart-mind, it will collapse." "*Qi* unites rightness (*yi*[a]) and the Way (*dao*). It will collapse if deprived of these" (*Mencius*, 2A:2). Supported by the accumulated righteousness, one can face a big army single-handedly without fear, and one's *qi*, supported by the moral rightness, can be so vast and unyielding that "it will fill the space between Heaven and Earth" (*Mencius*, 2A:2). It tells us that for Mencius and other Confucians, morality is neither a matter of decision making guided by some principles, nor is it simply a matter of acquiring mental/intellectual education. One needs to practice righteous deeds, and as one keeps practicing, one's *qi* gets purer and stronger.[6] Some later Confucians used *qi zhi* (temperament) or *qi xiang* (the *qi* outlook) to name this change. For example, Lu Xi Zhe says:

> A novice should pay attention to *qi xiang*. When *qi xiang* is good, everything will be smooth. One's *qi xiang* can be detected clearly just by looking at one's appearance, one's way of talking and acting – fast or slow, heavy or light. Not only does *qi xiang* differentiate morally superior and inferior persons, it determines whether one will have a successful or unsuccessful life, a long or short life (*Song Yuan Xue An – Ying Yang Xue An*).

Zhang Zai says: "the great benefit of learning is that it can change one's *qi zhi*" (*Song Yuan Xue An – Heng Qu Xue An*). Again, the learning that Zhang Zai refers to here is both intellectual and practical. Zhang Jiu Chen says, "Among the ancient sages, Mencius developed a unique theory of *qi* cultivation. It is superb, and his ideas show deep insight ... Those all came from his diligent and persistent daily practice, not something that one can gain by imitating the outward appearance" (*Song Yuan Xue An – Heng Pu Xue An*).

IV. CARE FOR OTHERS

Mencius' theory regarding *qi* has another point that calls for particular attention –- that the abundance of *yi*[a] can strengthen one's *qi* so much that one can go beyond being healthy in an ordinary sense. The *qi* nourished by *yi*[a] brings human power to its utmost potential. Such a person is capable of bringing great benefits to the entire world.

One might say, Confucius does not limit himself by making a person merely *capable* of benefiting the world; he teaches that one *ought* to benefit the world. In other words, his teachings are moral imperatives. Certainly Confucian virtues go far beyond the consideration of one's own health. Among his descriptions of what is *ren*, most have to do with one's being unselfish, and caring for others. Confucius' most famous description of *ren* is the "Golden Rule" – "Never impose on others what you yourself do not desire" (*Analects*, 12:2; 15:24). He also puts *ren* as "loving the people" (*Analects*, 12:12), and as *shu* –- "a benevolent man helps others to take their stand in so far as he himself wishes to take his stand, and gets others there in so far as he himself wishes to get there" (*Analects*, 6:30). He advocates many other virtues that essentially have to do with one's dealing with others, such as *zhong*[b] (doing one's best for someone), *xiao* (filial piety), *xin*[b] (being trustworthy in words), etc. But this does not mean that those virtues have no intrinsic connection with personal health. On the contrary, Confucians consider those virtues essential to and among the best ways of caring for one's own health.

Confucianism shares with other Chinese thought the kind of metaphysical outlook that F. S. C. Northrop calls a "continuum view of reality." According to this view, all within the universe are related and are parts of a cosmic flow. Distinctions or differentiations do not mean separation, and individuals *per se* cannot exist. In contrast with this view is the object-centered atomistic view of reality in the West, which is more centered on theoretical constructs, or abstract logical universals. Each object possesses its own essence, and can therefore be understood independently from its relation with others. In biology, the atomistic view has led to a theory that takes atoms, molecules, and cells as the final explanation, whereas the continuum view takes each organism as a link in a chain. Yale biophysicist Harold Morowitz quoted Northrop in a 1972 article stating that the steady move from the object-centered view toward the continuum view "represents what is perhaps one of the potentially most profound intellectual mergings of East and West" (Morowitz, 1972, p.46). In Confucianism, this continuum goes beyond the "natural realm" to the social. To Confucians, Heaven, Earth, and humanity form a trinity or ternion, where the "human" realm is obviously social since a human being inevitably lives in relation with others. For example, everyone is born a son or daughter, and is therefore born in a specific relation with his or her parents. One's relation with one's parents, with one's ruler, and with one's teacher are listed together with one's relation with Heaven and Earth as the five most basic relationships that one has to honor. These relations are so essential that they are the primary means by which an individual is defined. Just like a heart can function as a heart only within a body, and in relation with other parts of a body, a human can be a human only in relation with others.

From this metaphysical outlook, Confucians take human relationships very seriously in their dealing with personal health. They understand that a person's health is deeply affected by, and in turn affects, other people. For example, there is a reciprocal relationship between people: those who treat others with kindness and respect are more likely treated by others the same way. Therefore, treating others well is also caring for one's own well-being. I did not realize the profundity of this reciprocal relationship until I was struck by a statement made by Yan Xin, a grand *qigong* master from China. When he was asked about how to overcome being excessively nervous and fearful, he answered: "be more filial to your parents!" Here the space does not allow me to offer a detailed explanation of how this master provided more than enough to convince me, a philosopher who received a strict training in the empiricist/skeptic tradition, to believe that his instruction should be taken no less seriously than any instruction one can receive from a doctor. It is obvious that the master was not merely providing a sermon on morality. He was literally teaching a method or technique for a person to obtain a healthy state. From this there are a few important lessons:

First, it shows that the reciprocal relationship is not merely a mental reaction, whether intellectual or emotional. It is not a matter of affecting other people's rational decisions. In *qigong*, people use the phrase "*qi* field." When you treat others well, you are creating a harmonious and friendly *qi* field for yourself. As Song Jian and Yin Wen say, "the *qi* cannot be stopped by force, but can be pacified by virtue; it can not be retrieved by sound, but can be greeted by will. Respect it and do not lose it, that is to have complete virtue. When virtue is complete, wisdom is there and ten thousand things will all be obtainable" (*Guan Zi – Nei Ye Pian*). The similarity between this saying and Mencius' saying "The ten thousand things are all here at my disposal" (*wan wu jie bei yu wo*) (*Mencius*, 7A:4) is obvious. Tu Wei-ming is one of the few scholars today who have demonstrated a deep insight on this issue: "As a matter of fact in Mencius' theory of heart-mind ... flood-like *qi*, as both what fills the body and what fills the space between Heaven and Earth, is a real feeling that can be experienced by the person. Even his 'ten thousand things are all here at my disposal' is not a projection of subjective expectation, but an existential statement ... If we can cultivate our 'night *qi*' gradually up to a level that it becomes so vast that nothing can stop it, then filling the space between Heaven and Earth is just an inevitable result" (Tu, 1984). This explains why the Confucian program takes "*xiu shen*[a]" as a means for regulating the family, governing the state, and bringing peace to the world. The way it works is not merely that the sage ruler sets an example for others to follow, but that the moral person creates a magical power, or *qi* field, that literally affects others and, in turn, oneself.

Second, this example shows that Confucian moral virtues, and moral virtues in general, have an intrinsic relation with personal health. Confucian ethical teachings can be taken as techniques or recipes for obtaining, maintaining, and increasing personal health. By "intrinsic," I mean they do not merely regulate health care related behaviors (such as the principle "thou shall not lie" prohibits a doctor from lying to a patient), nor do they merely provide favorable conditions for health care (such as being *xiao* – filial – to one's parents provides the parents a condition of being cared for, or *ai* – love – results in actions of caring for young children, or *cheng* – sincerity – makes doctors honest and teaches them not to use patients as means for getting rich and famous, etc.), they are health care itself, like eating nutritious food, or doing physical exercises. In this sense, the practice of Confucian virtues is the practice of health care, and the possession of these virtues is the possession of health.[7]

If the intrinsic relation does exist, it should deeply affect our philosophical discussion regarding the fact-value dichotomy. How one ought to treat others will be an intrinsic part of the question about how one ought to treat oneself, which in turn is a matter of wanting to be healthy or not. It fuses the study of what ought to be the case and what is the case – technical questions like what will be the most effective methods for personal health care will be inseparable from ethical questions about what is moral. It may look like we are reducing the status of morality by associating it with personal health. This would seem contrary to the Kantian view that morality is a matter of "categorical imperatives" (do the right thing for the sake of its being right), not "hypothetical imperatives" (do the right thing only because it can bring about some desired end). Yet why do we have to have "either/or," and not "both" here? Morality as a technique might be of a special kind, such that when one uses it to obtain a reward, one must not do it out of the intention of attaining the reward – it works (or works best) only when one has no personal gain in mind. This formulation fits some religious teachings on morality. In Christianity, for example, good deeds may help to assure one's place in heaven, yet one must not do these deeds out of the motivation of going to heaven. In Buddhism, doing good deeds is a path to nirvana, yet one must not think of nirvana in order to reach it.

Third, this particular instruction implies that the *qigong* master has a special vision, and that it must be gained through a diligent practice and cultivation. How would an ordinary scholar associate fear with a lack of filial piety? It is very likely that ancient Confucians like Mencius had a similar vision, and statements like "a person with great virtue will surely have longevity" (*the Doctrine of the Mean*, ch.16) were made with that kind of vision in mind. To show that this is not incidental, I would like to quote a Jin Dynasty Daoist, Ge Hong (284-364), who was first a Confucian: "Those who want to be immortal must take *zhong*[b], *xiao*, *ren,* and *xin*[b] as fundamental. Those who search only formulas and skills without cultivating one's morality will not get a long life" (*Bao Pu Zi – Nei Pian*). The fact that this vision is gained through practice and not through theoretical speculation has tremendous philosophical significance. It tells us that moral knowledge is possible, but is gained not though theoretical discussion alone; it comes through practice and cultivation. This is not simply a place where "everyone is entitled to his or her own opinion." There are visions that are possible only to the cultivated mind. It tells us how mistaken contemporary moral philosophers have been when they take bioethics to be merely "applied ethics," as if one can work out some set of theoretical principles and then apply them straightforwardly to practice.

V. CONCLUDING REMARKS

This essay examined Confucianism through the perspective of health care. This does not mean that Confucianism ought simply to be reduced to a health care system. Rather, one can understand Confucian virtues as techniques for achieving optimal health. Confucianism is a system in which many distinctions are relative and almost non-existent, such as the distinction between health care and other types of well-being, whether intellectual or social, the distinction between the well-being of oneself and of others, the distinction between biological and psychological, moral and technical, learning what is good and learning how to be good, etc. The advantage of Confucianism is precisely in seeing the relativeness of all those distinctions, which reveals the connections between those realms. It is a health care system, but health care is not a matter of biology alone; it is a noble cause, an aesthetic ideal, and a never ending journey toward the highest perfection of a human being. It is a program of personal health care, but the persons are not isolated individuals who are bound together only by external contracts and protected by laws and moral rules for their rights. In this program, the perfection of one's own health is identical with one's striving for the perfection of others and the whole world.

Should a typical Confucian be characteristically healthy? Yes, but until the whole world is in harmony, it will not be possible for one to obtain fully one's ideal state of personal health.

Department of Philosophy
Grand Valley State University
Allendale, Michigan, USA

NOTES

* I would like to thank the editor of this volume for his helpful suggestions and clarifications. I also would like to thank my colleagues at the Department of Philosophy at Grand Valley State University for their thought-provoking inputs on the subject matter.

1 Though this kind of concept of health usually blurs the boundaries between personal health and economical, political, and social well-being, it has the advantage of allowing us to see more intrinsic connections between them.

2 Jiao is quoting *Chuang Tzu*, ch. 19, "*Da Sheng*."

3 "What I call 'human desires' are the sickness of the heart-mind. If one follows [them] one's heart-mind will be selfish and evil" (Zhu Xi, *Xin Chou Yan He Zou Za Er, Zhu Wen Gong*

Wen Ji, vol. 13.). "Many people are confused and blinded by the desires for material possession. That is the evil of the heart" (Zhu Xi, *Zhu Zi Yu Lei*, vol. 71.).

[4] The passages he quotes from the *Analects* are 7:29, 12:1, 15:4, 13:6, 2:1. See Fingarette, 1972, p.4.

[5] Confucius did not use the concept *qi* much. Other than the reference to "blood *qi*" which we have quoted in section II, we find that the *Book of Rites* contains some references to *qi*: "*qi* is what makes *shen*[b] magnificent" (*Book of Rites – Yue Ji Pian*). "When *qi* is strong, it transforms into *shen*[b]"(*Book of Rites – Ji Yi Pian*). Mencius made a significant contribution to the Confucian account of *qi*.

[6] One might ask "can a 'bad guy' have *qi*?" I think the Confucian would say "yes." For if *qi* is what pervades and animates the body, then every alive person has *qi*. While righteousness is one major source of *qi*, there are other sources. Even having some money can make one's *qi* a little stronger. When one is accompanied by a crowd of others with the same will, one's *qi* can also increase. Yet, the Confucian would emphasize that without righteousness the *qi* cannot reach a really high level and cannot last for long.

[7] I do not mean to exclude extrinsic connections. Confucian ethics does have an extrinsic connection with personal health. Among the traditional rites that Confucians advocate, many are ethical codes that provide extrinsic conditions for health. For example, "When a ruler drinks medicine for sickness, a subject should taste it first, when a parent drinks medicine for sickness, a son should taste it first" (*Book of Rites*). The purpose for tasting medicine is to test whether it is poisonous, since traditional medicines were made directly from herbs.

REFERENCES

Chuang Tzu: 1986, *Chuang Tzu* in *Zhu Zi Ji Chen*, vol. 3, Shanghai Shu Yu, Shanghai.

Confucius: 1986, *The Analects*, in *Zhu Zi Ji Chen*, vol. 1, Shanghai Shu Ju, Shanghai; and Arthur Waley, trans.: 1938, *The Analects of Confucius*, New York: Vintage Books.

—— 1967, *Li Chi* (*The Book of Rites*), in James Legge (trans.), New York: University Books.

—— 1893, *The Doctrine of the Mean* (*Zhong Yong*), in James Legge (trans.), *The Chinese Classics*, with Chinese text, vol. 1, Clarendon Press, Oxford.

—— 1893, *The Great Learning (Da Xue)*, in James Legge (trans.), *The Chinese Classics*, with Chinese text, vol. 1, Clarendon Press, Oxford.

Dong Zhong Shu: 1936, *Chun Qiu Fan Lu*, in *Si Bu Bei Yao*, Zhong Hua (Chung-hua), Shanghai.

Fingarette, Herbert: 1972, *Confucius, The Secular as Sacred*, Harper & Row Publishers, New York.

Ge Hong: 1973, *Bao Pu Zi*, in *Wan You Wen Ku*, Commercial Press, Shanghai.

Huang Di Nei Jing: 1976, Xian Zhi Chu Ban She, Taipei.

Ling Shu – Zhong Shi.

Su Wen – Ba Zheng Shen[b] *Ming Lun Pian.*

Su Wen – Shang Gu Tian Zhen Lun.

Su Wen – Si Qi Tiao Shen[b] *Da Lun.*

Su Wen – Tiao Jing.

Jiao Xun: 1987, *Meng Zi Zheng Yi*, Zhong Hua Shu Ju Press, Beijing.

Lu Xi Zhe: 1934, *Song Yuan Xue An – Ying Yang Xue An*, reprint; Commercial Press, Shanghai.

Mencius: 1970, *The Works of Mencius*, in James Legge (trans. with Chinese text), Dover Publications, Inc., New York.

Morowitz, Harold J.: 1989, 'Biology of a Cosmological Science,' in J. Baird Callicott and Roger T. Ames (eds.), *Nature in Asian Traditions of Thought, Essays in Environmental Philosophy*, SUNY Press, Albany, New York.

Shi Qi, Lu Mingfang: 1990, *A Grand Compilation of Practical Chinese Health Preservation Theories and Prescriptions (Shi Yong Zhong Guo Yang Sheng Quan Shu),* Xue Lin Press, Shanghai.

Song Jian and Yin Wen: 1986, *Guan Zi – Nei Ye Pian*, in *Zhu Zi Ji Chen*, vol. 5, Shanghai Shu Ju, Shanghai.

Sun Si Miao: *Bei Ji Qian Jin Yao Fang*.

Tang Zhong You: 1934, *Song Yuan Xue An – Shuo Zhai Xue An*, reprint, Commercial Press, Shanghai.

Tu Wei-Ming: 1984, 'Viewing Confucian Theory of Human Being from Four Levels – Shen[a], Xin[a], Lin and Shen[b],' *Ming Bao Monthly*, November, 1984.

—— 1989, *Centrality and Commonality: An Essay on Confucian Religiousness*, SUNY Press, Albany, New York.

Wang Yang Ming: 1933, *Wang Wen Cheng Gong Quan Shu*, Commercial Press, Shanghai.

Xun Zi: 1935, *Xun Zi*, World Books Co., Shanghai.

Zhang Zai: 1934, *Song Yuan Xue An – Heng Qu Xue An*, reprint, Commercial Press, Shanghai.

Zhu Xi: 1934, *Song Yuan Xue An – Hui Weng Xue An Shang*, reprint, Commercial Press, Shanghai.

—— 1966, *Xin Chou Yan He Zou Zo Er*, in *Zhu Zi Da Quan*, Chung Hua Shu Ju, Taipei.

—— 1983, *Zhong Yong Zhang Ju Zhu,* in *Si Shu Zhang Ju Ji Zhu*, Zhong Hua Shu Ju, Beijing.

—— 1983, *Zhu Wen Gong Wen Ji*.

—— 1973, *Zhu Zi Yu Lei*, Cheng-Chung Shu Chu, Taipei.

ELLEN Y. ZHANG

THE NEO-CONFUCIAN CONCEPT OF BODY AND ITS ETHICAL SENSIBILITY

"How could we abandon our life energy and physical bodies and look somewhere else for the principles of righteousness?"

Feng Congwu (1556-1627)

In the *Great Learning*, one of the Confucian classics, we read, "from the son of the Heaven down to common people, all should take *xiu shen* as their ultimate concern" (*Da Xue*, Preface). *Xiu shen* is also one of the key terms in a cluster of Neo-Confucian language.[1] Throughout history, the term *xiu shen* has been interpreted as "self-cultivation," the "cultivation of person (*ren*)," the "cultivation of morality (*de*)," and the "cultivation of the mind (*xin*[a]) and spirit (*shen*[b])." In Chinese language, *xiu shen* means literally "the cultivation (*xiu*) of the human body (*shen*[a])." Here, we see several possible meanings imbedded in the concept of self-cultivation. This lack of a fixed, univocal conceptual terminology of *xiu shen* may sound quite odd to many Westerners who are trained in the line of the Western philosophy, yet it is the polyvalence in words and concepts that well characterizes the unique nature of the classical Chinese philosophy.

Yet I am by no means suggesting that body is mind or vice versa in traditional Chinese thought because I am fully aware of a hermeneutic sensibility when we attempt to pin down precise English terms for the Chinese word "body," which in Chinese can be *shen*[a], *ti*[a], and *xing*. The difficulties of representation have often been coupled when we try to employ Western philosophical language such as the body-mind unity (or dichotomy) to explicate Neo-Confucianism where a pure clarification of concepts such as body and mind is not often an issue. Nevertheless, we can still explore the Neo-Confucian mode of ethical reasoning in light of the conception of body and its moral implications, and examine how through the notion of *xiu shen* the Neo-Confucians maintain the endowment of body as an indispensable qualification for personal and social existence. Also, we can examine how the idea of body and the perception of life-experience embedded in body are closely related to the fundamental Confucian teaching that "the Dao (the Way, the Virtue) is not far off from humanity (*dao bu yuan ren*)."

Ruiping Fan (ed.), Confucian Bioethics, 45–65.

As a syncretism of traditional Confucianism (that is, the teachings of Confucius and Mencius), Daoism, and Buddhism, Neo-Confucianism (despite a divergence of teachings among Neo-Confucian philosophers themselves) has perpetuated the basic Confucian tradition with its passionate moral inquiry and a strong sense of social obligations. But at the same time, Neo-Confucian philosophers also search for metaphysical transcendence in terms of self, society, Heaven, and Principle. Through the tenet of *xiu shen*, they attempt to trespass the boundaries between the Nature of Heaven and the principle of life, between the noumenal One and the phenomenal many, between eternity and temporality, between "what is above the shape" and "what is within the shape," and between the process of humanization and the journey of spiritual realization.

I. THE ONTO-COSMOLOGICAL BODY AND THE PERSONAL BODY

For most Confucians the human body, like the body of any other being in the world, should not be seen as an autonomous entity since the human being is part of the natural world, which comprises the totality of Heaven-Earth (*tian di*), humanity (*ren*), and myriad things (*wan wu*). This integrated **one** body is also called the Dao, which embraces both ontological and ethical implications. According to the Confucian onto-cosmology, the human body in its constitution is inseparable from the body of the cosmos, for the human body shares the flowing course of nature through the life-energy (*qi*) from Heaven. More specifically, the human being is an embodiment of the basic five elements (*wu xing*); that is, water, fire, earth, metal, and wood. These five elements have, in Chinese tradition, become a cultural mythos infused into everything in Chinese thought, from onto-cosmology to the explanation of various physiological, pathological, social, and moral phenomena. It is believed, for instance, that when the five elements are out of balance in the human body, one experiences sickness; when the five elements are out of balance in the human mind, one experiences moral decay; and when the five elements are out of balance in society, the society experiences chaos. For Confucian thinkers there is an internal mutual connection between the onto-cosmological body and the physical bodies of both human beings and human activities. In other words, the onto-cosmological body can constrain the way the physical body is perceived, and at the same time, the physical body can constantly modify the onto-cosmological body.

One of the best examples of the extension from onto-cosmology to human activities can be seen in the correspondent relations indicated in the following table where we can find a plethora of physiological relationships between the five elements (*wu xing*) and other things in the world. They cover the full range from the five directions (*wu xiang*) to the five colors (*wu se*), the five flavors (*wu wei*), the five sense organs (wu guan), the five body parts (wu ti), the five internal organs (*wu zang*, or *wu zhu*), the five grains (*wu gu*), the five inclinations (*wu zhi*), the five sounds (*wu sheng*), the five classics (*wu jing*), the five constant virtues (*wu chang*), and the five constant relationships (*wu lun*).

five elements	wood	metal	fire	water	earth
five directions	east	south	west	north	center
five colors	blue	yellow	red	white	black
five flavors	sweet	sour	bitter	pungent	salty
five sense organs	ears	eyes	lips	nose	tongue
five body parts	head	left arm	right arm	left leg	right leg
five internal organs	heart	liver	spleen	lungs	kidney
five grains	rice	millet	sorghum	wheat	beans
five inclinations	anger	worry	pleasure	fear	thinking
five sounds	scream	laugh	cry	sing	moan
five Classics	*Book of Songs*	*Book of History*	*Book of Change*	*Book of Rites*	*Spring & Autumn*
five virtues	humanity	righteousness	loyalty	wisdom	sincerity
five relationships	ruler–ministers	father–son	husband–wife	the old–the young	friend–friend

Such a holistic view is also accepted by Neo-Confucian philosophers who regard the human body as both a biological organism and a life-energy by which man bears the principle of Heaven-Earth and links himself to other things in the world. The claim that "Heaven-Earth and I are in the same body, and the myriad thing and I are in oneness" *(tain di yu wo tong ti, wan wu yu wo wei yi*) is a motto for many Neo-Confucians of the *lixue* School. Shao Yong (1011-77), for example, believes that the human body is that of the universe (Heaven-Earth, *qian-kun*), hence we should regard the universe as our own parents. "*Qian* is my father and *kun* is my mother," says Shao Yong, "my mind is *qian* and my body is *kun*" (*Huang chi ching shih*, 1962). Zhang Zai (1020-77) also maintains that Heaven is man's father, and Earth is man's mother. Cheng Yi (1033-1107), in the same manner, claims, "Taking something close like our bodies, the myriad things are all present and complete" (*Er Cheng Ji*, 1981, p.167). Wang Yang-ming (1472-1529) a Neo-Confucian of the Ming Dynasty and an admirer of Shao Yong, holds the view that a man with virtues regards Heaven and Earth and the myriad things as one body.[2] It follows that self-cultivation is, for Neo-Confucians, not only to realize Heaven's humanity in oneself, but also to extend it to actual human circumstances, as what is said in the *Great Learning*, "To regulate the state, one needs to regulate one's family; To regulate one's family, one needs self-cultivation" (*Da Xue*, Preface).

This integrated oneness of the Heaven-Earth, humanity, and oneself is further reinforced when Neo-Confucians employ the conception of the "Supreme Ultimate" (*tai ji*) to the classical Confucian understanding of Heaven and Nature. Zhou Dun-yi (1017-73) in his *tai ji* schema places the human being in the pivotal position of the system of *yin* and *yang*, and maintains the internal unity between the forces of the natural world of Heaven-Earth and the forces of human beings. Furthermore, he sees the emanation of the life-energy in the human body as the manifestation of the "Supreme Ultimate" that calls for recognition in the living ambiance. It follows that for Zhou Dun-yi and other Neo-Confucians, the "Supreme Ultimate" becomes an onto-existential energy-resource upon which one understands the world and humanity. The "Supreme Ultimate" also suggests the internal link between one's body and one's mind. In other words, a person is a **living** body that encompasses one's whole being, including human desires in the world. According to Confucianism, a mind with no body is not real and a body with no mind is just a "walking corpse and running flesh." Therefore, in order to understand human nature and its relation to other beings, we should approach the human body holistically.

It is in the connection of body and mind as a whole being that Neo-Confucian philosophers see the role of "self" in building an ethical system. For Neo-Confucians the self is psycho-physical and includes a body-image (a particular personal body vis-à-vis other bodies) as well as a spiritual self that feels and wills. Accordingly, *shen* is a "body-person," or a "personal body" that not only responds to the myriad things but also contemplates about itself and its relation to others.[3] Thus self-awareness, self-knowledge, and self-examination become an inevitable stage in the process of *xiu shen*. Shao Yong, for example, employs the notion of "self" (*shen*[a], or *wo*) to his idea of "reflective perception" (*fan guan*) by which one realizes the inner self through an understanding of the totality of the self, Heaven-Earth, and myriad things. He sees this understanding of the self and its relation to things as "self-awareness" (*guan xing*). Shao Yong divides the process of understanding into three steps: 1) sensing, 2) thinking, and 3) knowing. In the third stage, the distinction between subjectivity and objectivity no longer exists.[4] Here, it should be noted that self-awareness for Neo-Confucians differs from the Cartesian *ego cogito* since the self involves the entire body and mind of a person, and self-cultivation involves work on both self-knowledge (i.e. the particular self) and no-self (the universal self).

Zhu Xi's concept of "reflective examination" (*xing cha*) suggests similar meanings as does Shao Yong's when he regards thinking as a form of reflective examination of the self. For Zhu, one should keep examining the self in order to obtain the Principle of Nature and enter the sage-hood because to become a person with humanity (*ren*, that is, a high moral principle) requires the process of transforming the physical self by developing a heightened awareness of the self-body (feelings and intentions). Zhu also maintains that self-actualization or self-exemplification of humanity is a foundation for a moral society as a whole. He argues, "whenever and wherever humanity flows and operates, righteousness will fully be righteousness and propriety and wisdom will fully be propriety and wisdom" (Chan, 1963, pp. 632-33). As such, the self in Neo-Confucian teachings does not necessarily mean to be isolated or alienated from society. Rather, it is self-cultivation *qua* cultivating others because every human being has a potential to be a person with virtue. For Neo-Confucians, there is no distinction between the internal sage-hood and external kingship. This is why Neo-Confucians maintain that the purpose of cultivation is to bring virtue back to society, back to what Tu Wei-ming calls "the community of the selfhood" (Tu, 1979, p.13). A cultivated person, then, is an exemplar of human virtue who can exert a transforming influence on society.

One point needs elaboration. *Xiu shen* in Neo-Confucian teachings conveys a notion of self-power vis-à-vis other power in other traditions. Neo-Confucians insist that the principle of Heaven-Nature exists in human mind, and that the Dao is identical with the nature of man. So there is essentially no distinction between sage (*sheng*) and man (*ren*). Through self-efforts and self-power, anyone can enter into the inner sagehood. This perfectibility of human beings through self-effort marks the fundamental difference between Confucianism and other traditions that posit the other-power through God, or other supernatural agents. The non-theistic view of Confucian philosophy determines its emphasis on the natural order of nature rather than the grace of God. In the Heaven-Man unity, a Neo-Confucian sees the power in everything, including in oneself, while in theistic views one generally sees the power in nothing but in God. This fundamental difference between Heaven and God is difficult to reconcile.

II. SELF-CULTIVATION: THE HARMONIOUS FIT BETWEEN *SHEN* AND *XIN*

In the Western intellectual tradition, the word "body" as associated with "container" images is derived from the Anglo-Saxon *bodig*. The Old High German word *botah* means a cask, a brewing tub, or a vat. The imagery suggested by the word basically indicates the material being, that is, the corporeality of body and, as such, body is viewed as the opposition to the word "soul" or "spirit." This interpretation of the human body leads to a dualistic concept of a corporeal and corruptible body that needs to be "ensouled," or "resurrected" in the Western theological/philosophical discourse.

The word *Shen* in Chinese philosophy, on the other hand, is often used not simply as the human physique in its corporeality; rather, it is used to refer to an entire physic-psychic person. As such *shen* is frequently used with the term *xin*[a], which, different from the word "mind" employed in the Western philosophical tradition, often entails both the meaning of physicality (because *xin*[a] basically means "heart") and the meaning of spirituality. It follows that a human being is regarded as an integrated whole (*shen*[a] and *xin*[a]) rather than something bifurcated into mind and body.

In the Chinese tradition, though we have such expressions as "to sacrifice body for the sake of benevolence" (*sha shen qiu ren*), body as a whole is not seen as an entity that is in contrast to soul or spirit. With its complex and organic life-energy, *shen*[a] embraces both a material force and a spiritual force. Accordingly, we find that *shen*[a] is sometimes defined with its other two homophones: *shen*[a] with a hand radical means "to extend," or "to stretch out," and *shen*[a] with a heart-mind radical means the inner spirit. The extension here suggests a correlative relationship between *shen*[a] (the physical) and *xin*[a] (the spiritual). Therefore, we have the Neo-Confucian argument that by guiding "pure breath" (*qing qi*) through the body, a person brings in pure mind (*qing xin*). This extension of meanings from the physical to the ethical can be also observed in many other expressions in Neo-Confucian terminology. For instance, *dao* means both a physical path and moral virtue; *ti*[a] signifies both shape and substance; *li*[a] indicates both the texture and principle of things; and *feng gu* suggests both a natural phenomenon and a moral virtue. Such analogies between the physical and the spiritual are abundant in Chinese philosophical discourse.

Since body and mind (or *da ti* and *xiao ti* in the Mencius' terms) are not regarded as separate entities in any essential way, the negative connotations toward the body are rarely present in Confucianism and Neo-Confucianism. To most Confucians, both body and mind possess material forces and the inner energy, and there is a close connection between bodily life and ethical life. Therefore, in Confucian teachings, the "cultivation of body" (*xiu shen*) and the "rectification of mind" (*yang xing*) are often used interchangeably. Furthermore, the unity of body and mind is expressed by the notion of a harmonious fit (*he*), or the balance of the *shen-xin*. The idea of *he* indicates a harmonious balance between the human body and the human mind, between one's medical condition and one's moral conduct. Hence, Neo-Confucians contend that disrupted harmony is the fundamental cause of disease in both one's physical endowment and moral character. Cheng Hao (1032-85), for instance, relates the paralysis of hands and feet (*shou zhu*) to the impairment of the mind because he believes that when one is paralyzed, the life-energy is blocked, which in turn can block the moral energy.[5]

For Neo-Confucians the activities of learning (*xue*) then should not be limited to a pure mental training, or confined to what in the West is called "mind," since the activities of learning are meant to be both spiritual and physical. Therefore, despite their differences in many other philosophical arguments, both Zhu Xi and Wang Yang-ming regard learning as that which involves the entire body and mind. Zhu Xi sees learning as a mediating method that links body and mind and, as such, "in reading books, one's body and mind must enter inside each paragraph" (Smith, *et al.*, 1990, p181). In the same fascicle, Wang Yang-ming insists that the "learning of body and mind" requires a correlative operation of *shen*[a] and *xin*[a] in the learning process. For Neo-Confucians, the sage is a person whose body and mind are both composed of clear energy (*qi*). To become a sage, then, is a matter of cultivating one's energy. Hence *xiu shen* is both a physical and a spiritual training in the process of cultivation. The cultivation of body and mind can be shown in the following diagram:

body (*shen*[a])
the harmonious fit <————————> the moral balance
mind (*xin*[a])

In doing so, Neo-Confucians have established a profound relation between *shen*[a] and *xin*[a], a relation with both physical and mental/moral dimensions. As one's mind and body become a integrated oneness, the internal process of self-cultivation also becomes one and is integrated with the spirit-matter of heaven and earth. This oneness of body and mind is, according to Neo-Confucians, back to the true nature of human beings. Therefore, for Neo-Confucians the balance of *shen-xin* or the harmonious fit, include a process of dieting, internal cleansing, exercising, and guided meditating. The cultivation of the body is required in order to reach the "spiritual goal," that is, the moral balance. In Neo-Confucianism the practice of "quiet sitting" (*jing zuo*) or seated meditation, for example, offers a paradigmatic interpretation of the role of the body in our understanding of the body-mind coagulation In Neo-Confucian discourse. The relationships between the body and the mind, between the exterior world and the interior world can be established in a particular body posture. The whole process of closing one's eyes (*ming mu*), isolating oneself (*bi guan*), cross-legged sitting (*fu zuo*), breath-control (*tiao xi*) is designed to control one's mind (including controlling emotions) through controlling one's body *via* breathing exercises.[6] The purpose of breathing exercises is to "empty one's mind" so that the clear energy can pervade the whole body from the head down to the toes and finger-tips. This notion of emptying one's mind resembles the Daoist theory of sitting and forgetting, which also aims at a correlative relationship between the movements of the body and those of the mind. Zhuang Zi, for instance, contends that the mind of a perfect man "is like a mirror."[7] Yet "quiet sitting" in Zhuangzi somehow suggests a kind of detachment from social and ethical obligations, whereas "quiet sitting" in Neo-Confucianism is more ethically oriented.

It should be noted that Neo-Confucianism is often regarded as a form of rationalism because of its critique of human emotionality (*qing*[a]) and self-desire (*si yu*).[8] Excessive emotions and self-desire, as Neo-Confucians see them, obscure one's mind and lead to immoral conduct. Zhang Zai, for instance, argues that "He who understands virtue allows (his body) to have sufficiency of (drinking and eating) and no more. He neither enchains his mind with sensual desire, injures the great with what is small, nor destroys the root with what is peripheral" (Fung, 1953, p. 490). However, the harmonious fit of body and mind is meant to be moderate, rather than ascetic, because for most Neo-Confucian thinkers *xiu shen* is not a cultivation of virtue at the expense of man's basic physical nature. In the Confucian tradition, the physiological needs and desires of man – which include eating and drinking and sexual activity – have never been denied. Nevertheless, Confucians do believe that by virtue of cultivating one's mind, one's body can also be properly nourished. They contend that a moral person is able to prevent his inner spirit (*shen*[b]) from scattering and, as such, keeps his body intact. As Mencius puts it, "Virtue fosters the body" (*de run shen*, *Da Xue*, 6).

The Neo-Confucian concepts of equilibrium (*zhong*) and harmony (*he*) can help us to understand their critique of emotions and self-desires. Neo-Confucians do not deny the human emotions *per se*, merely inappropriate emotions; that is, the excess of desires and passions. Their critique is in the similar vein of Laozi's rejection of an excessive indulgence in the five colors, the five sounds, the five tastes, and an excess of hunting (*Laozi*, 12). According to Neo-Confucians, the harmonious fit requires balance: It is a middle-way of neither deficiency nor abundance. Here, the notion of *zhong he* in Neo-Confucianism is related to the idea of the mean (*Zhong yong*) presented in *the Doctrine of the Mean* in classical Confucianism. *Zhong yong* signifies the perfect equilibrium and harmony at the plane of both the physical and moral well-being. Therefore, Confucius says, "The superior man (*jun zi*) exemplifies the virtue of the means, the inferior man (*xiao ren*) acts contrary to the virtue of means" (Chan, 1963, p.98). Self-cultivation, therefore, should be the cultivation of both bodily and spiritual strength so one can find the middle path, without inclining to either extreme position.[9]

It is important, here, to remember that equilibrium and harmony (*zhong he*) are crucial concepts in Neo-Confucianism to which Zhu Xi attaches great importance. For Zhu, *zhong he* functions as both a critique of the excessiveness of passions and desires on the one hand and the Buddhist asceticism on the other. Buddhism, as Zhu understands it, renounces the family and, as such, renounces social responsibility. The notion of *zhong he* also leads to Zhu's critique of the excessiveness of human emotions (*qing*[a]), that is, the feelings of pleasure, anger, sorrow, and joy (Chan, 1963, p.601). Here, despite the fact that Zhu Xi is often labeled as a rationalist, like most other Neo-Confucian thinkers, he believes that there is nothing wrong with emotions *per se*; what matters is the excessive use of emotions, which turns out likely to interfere with one's practice of learning. It is Zhu's belief that the excess of emotion will decay the mind just as the overindulgence in meat, wine, and grains will clog up and decay the body.

On the other hand, the harmonious fit in terms of equilibrium and harmony also entails ethical embellishment regarding the differentiation between right and wrong. For a cultivated person, the ethical judgment is not a matter of correspondence between observation and things being observed, but a matter of coherence and harmony among things. It is in this sense that we say the Neo-Confucian concept of righteousness (*yi*[a]), or moral rightness, is "fitting" or "suitable" rather than "true" or "false." What a cultivated person seeks is what is suitable in a given situation. And as such he is fully aware of the possibility of flexibility and contingency in any moral decision, and he is capable of keeping a balance between unwarrantable skepticism and all-embracing dogmatic certainty.

III. THE TEMPORALITY OF HUMAN BODY

Though the awareness of death is not absent in Neo-Confucianism, death and the life after death have never been central themes in Neo-Confucian discourse. In fact, there is a lack of any form of metaphysical account of death or life after death in Neo-Confucian philosophy. Zhang Zai's well-known statement that "in life I shall serve unresistingly, and when death comes, I shall be at peace" (Fung, 1953 p.497) typically represents the Confucian attitude toward life and the temporality of the human body.

The most important question for a Neo-Confucian thinker is: How can we live our body in our current and everyday mode of existence?

In fact, the "temporality" of the human body is a relative term for most Chinese. According to Chinese belief, the human body is produced by Heaven through Heaven's life-energy (*qi*) no matter whether the body is condensed or dispersed, aggregated or separated. "Condensed, (the *qi*) forms my body; dispersed, it still forms my body," therefore, "life does not entail gain nor death losses" (Fung, 1953, p.497). Such a view of the body also leads to a rejection of the dissection of the human body in the traditional Chinese society. Nevertheless, there is a sense of the temporality of human life in Confucianism, and the temporality of human existence is explicated through an emphasis on the "now" and "here," that is, the body's existence in the world. Like Confucius who was not interested in God or spirits in general, Neo-Confucians are also not very attracted to the idea of "ghosts and spirits" (*gui shen*). In his defense of Confucianism, Zhu Xi, for instance, repeats what Confucius says, "People should be properly reverent towards ghosts and spirits, and yet keep them at a distance" (*Lun Yu* 6:20). Consequently, Zhu criticizes those who believe in and serve the Buddha since they, as Zhu sees it, fail to keep the spirit at a distance.

Very often, the temporality of the body is expressed by Confucians in terms of the phases of season. Tu Wei-ming points out, the change of the body as budding (spring), growing (summer), withholding (autumn), and preserving (winter) "symbolizes longevity as the pattern of human flourishing which honors the natural process of maturation rather than the artificial imposition of any discrete (for example youth) as an abstract ideal" (Tu, 1992, p.89). In the same vein, Confucians also understand the alternation of life and death as phases of the incessantly producing process of nature. Shao Yong, who regards death as "hiding away" (*cang*), contends that the life cycle has four stages including birth, growth, maturation and death. Therefore, "when one is born, one's nature is of heaven. When one is completed, one's form is of earth ... this is the dao of change" (Birdwhistell, 1989, p.117). This cyclical view of life and death resembles of the Daoist interpretation of the life circle where we have: 1. receiving breath, 2. being in the womb, 3. being nourished, 4. birth, 5. being bathed, 6. assuming cap and girdle, or puberty, 7. becoming an official, 8. flourishing, 9. weakening, 10. sickness, 11. death, and 12. burial.[10] This interpretation of the life cycle is basically accepted by Neo-Confucians. For them the transformation of life, that is, "becoming and perishing," characterizes human life and existence, therefore, a person should not worry too much about death.

Though Confucians speak about "nourishing the life" *(yang sheng)* to prevent one from sickness and un-natural death, they do not advocate extreme means intended to forestall old age. When death comes, one should regard death calmly, as a natural event, especially when it is a natural death (*shou zhong zheng qin*). It follows that Neo-Confucian thinkers refute both the Buddhist concept of immortality *via* Nirvana and the Daoist practice of immortality. The former speaks of life (generation) and death (extinction) in terms of "transmigration" (i.e. from human to animal to other forms of life), and the latter looks for immortality (*chang seng bu si*) *via* extraordinary medical means. For Neo-Confucians, what matters most is not whether one lives a long life, but whether one lives a meaningful life, that is, a life with virtue and dignity. Therefore, Zhang Zai tells us, "Perform each day the duties belonging to that day, serene in the consciousness that the coming of death merely means our returns to the Great Void from which we came" (Fung, 1953, p.497). This traditional Confucian ethos has remained basically unchanged throughout history.

Strictly speaking, the goal of *xiu shen* is neither immortality in a pure physical sense, nor in the sense of the Platonic immortality of the soul. Rather, the goal of *xiu shen* is to immerse the human mind-body into the mind-body of Heaven-Earth, where one's inner energy can be transformed into perceptions and values. The nourishment of the body and the cultivation of the mind are interdependent facets of a single activity rather than two separate concepts external to each other. Thus, *xiu shen* is an activity deeply rooted in the human world as an embodiment or a transformation of true humanity expressed by a fulfillment of the duties of man and the accomplishments in this world.

According to Confucianism, a person (alive or dead) is always connected to other things in the world. Therefore, when Confucians do speak of immortality, their conception of immortality is associated with blood (producing offspring) or with virtue. The former refers to the importance of marriage and reproduction, whereas the latter refers to importance of moral character exemplified by those who have shown virtue or those who perform great virtuous acts. Generally speaking, Confucians are not too concerned with any quest for a life beyond this world. Compared with the Buddhist conception of immortality, the

Confucian understanding of immortality is ethically oriented rather than metaphysically oriented.

IV. BODY AS A LIVING EXPERIENCE

According to Neo-Confucian teachings, the body is not only a physical entity where the mind dwells but also brings the Dao into practice (*xing qi dao*) by means of an embodiment of the true humanity (*ti ren*). The notion of practice (*xing*), then, becomes a crucial theme in the Neo-Confucian philosophical tradition. As such, we can find a plethora of expressions connected with the body and bodily actions, such as *ti hui* (to experience through body), *ti cha* (to observe through the body), *ti yan* (to verify through the body), *ti ren* (to comprehend through the body) and *shen ti li xing* (to practice through the body). Hence such activities as experiencing, observing, and understanding are not limited to being purely mental activities. In other words, for Neo-Confucians interiorization is dependent upon exteriorization, and morality, which includes concrete humanity, righteousness, propriety, and wisdom, is essentially moral practice *(de xing*). The Dao in Confucianism, therefore, is no longer an abstract concept, but a process of acting and performing. Accordingly, *xiu shen* means that only through practice can one put into effect one's knowledge, for cultivation is not simply a matter of spiritual nourishment but a behavior, a conduct, and an action.

Zhu Xi has observed that "in reading books, instead of simply looking for the patterns of the Dao (*dao-li*) on the page, one should come back and investigate one's own self...Knowledge and practice require each other... The effort of knowledge and practice should be exerted to the utmost" *(Chu-tzu Ch'uan-shu* 6:18b; 3:8a; 3:8b). For Zhu, Principle (*li*[a]) – a concept he sometimes employs to replace the *dao* or *dao-li* in classical Confucianism – should be the unity of the moral patterns of Heaven and the moral characteristics of human beings, and it is demonstrated through practice qua body experiences, that is, physical involvement. Zhu maintains that learning is bodily experiencing that requires participation in the phenomenal and the concrete (*shen ti li xing*). Only in such a way can the effort of both knowledge and practice be exerted to the utmost.

In Confucianism, we have the argument that "filial piety and brotherly respect are the root of humanity" (*Lun Yu,* 1:2). Hence, humanity in Confucian teachings is not just a principle that one tries to learn and understand, but concrete human relationships that one has to experience, including the relationship between ruler and ministers, between father and son, between husband and wife, between old and young, and between friend and friend (Legge, 1970, p.252). It is quite natural, then, for most Neo-Confucians to maintain a view regarding the inseparable relationship between learning and ethical conduct. For them morality is not simply a form of language played out in terms of pure theories; rather it is a process of an actualization of those theories through proper conduct of humanity. As Tu Wei-ming points out, the Confucian concept of *cheng* conveys not only the idea of "sincerity," but also the connotation of "completion," "actualization," and "perfection" (Tu, 1978, pp. 97-99). The ethical conduct itself is the embodiment of the highest idea of sagehood. Here again, we find a strong sense of performance and action.

This emphasis on morality with practical affairs also leads Neo-Confucian thinkers to argue that the body is both substance (*ti*) and function (*yong*).[11] From Zhu Xi's "investigating things"(*ge wu*) to Wang Yang-ming's "unity of knowledge and practice" (*zhi xing he yi*), we find a dynamic process of transformation in which the activity of self-cultivation becomes real experience rather than an abstract concept. The linkage between knowledge and experience characterizes the process of humanization. Therefore, for most Neo-Confucian philosophers, antiquity should not be seen as providing closure for the teaching of human virtue and values. Sages like Confucius and Mencius have set good moral examples, yet their teachings should not be taken dogmatically.

With respect to the role of the body in performance and action, Roger T. Ames has made an interesting point when he delineates the element of physicality in the Confucian ritual practice. Ames considers ritual practice or propriety (*li*) as "an action embodiment or formalization of meaning and value that accumulates to constitute a cultural tradition."[12] This observation corresponds to Tu Wei-ming's claim that *li*[a] in Neo-Confucianism is the "externalization" of morality (*ren* or *de*) in concrete social circumstances. Therefore, through the "performance" of the body, one does not simply learn the past (the "heuristic or pedagogical function" of *li*) but transforms reality via *xiu shen,* that is, by cultivating the original human nature. Here, by seeing *li* as a "physical rendering" of moral meanings or a "moral process" of "actualizing *ren*" rather than a subordination to dogmatic doctrines, both Tu and Ames attempt to resolve the tension between the ritual practice and self-cultivation in Confucianism in general and Neo-Confucianism in particular.

In the Neo-Confucian vocabulary, the Dao, Heaven, the Supreme Ultimate, Principle, the Heavenly Principle, and Nature basically entail the same meaning, which, to most Neo-Confucians, represents a "fullness of being and goodness." The oneness of Heaven and man (*tian ren he yi*), therefore, is both an onto-cosmological argument and an ethical experience. From a Neo-Confucian perspective, to approach the Dao, to understand the Heavenly Principle, or to enter the sage-hood is to know how to live in the world, and how to effect harmony in human activities. The Dao or Heaven, then, is not only the undifferentiated world of the transcendent but also the differentiated world of concrete life itself. Hence the oneness of the Dao/Heaven, as Neo-Confucians see it, opens up to a realm that is not only onto-cosmological but also ethical. Therefore, we have:

			(cosmic order)
	Dao of Heaven	→	onto-cosmological
The embodiment ←→			
of the Dao ←→			↑↓
	dao(s) of man	→	ethical
			(moral order)

As this diagram illustrates, the Dao is both the principle and the actualization of this principle. This explains why Neo-Confucians hold the view that the Dao is not far off; it is right here in one's body. The embodiment of the Dao (*daoti*) in Neo-Confucianism, then, indicates a mutual transformation between the corporeality of the dao and the essence of the Dao.

V. ABOVE THE SHAPE AND WITHIN THE SHAPE: THE BODY-MIND POLARITY

Another word used to represent the notion of body is *xing*. This also means the "form," "shape," or "appearance." Neo-Confucians also equate *xing* with the concept of *qi* (which means instrument, not the same *qi* as life-energy). Both *xing* and *qi* refer to the physical things in the world. But how is this form or shape as body connected to the Principle of Nature as mind? Cheng Hao, for example, maintains that there is no principle apart from actual things and, as such, there is no distinction between the metaphysical and the physical. "Fundamentally... what is within the shape are nothing but the Dao" (Fung, 1953, p. 511). While speaking about *xiu shen*, Cheng Hao argues, "The student need not seek afar. Let him take what is near his own person... Heaven and man are one in that they both have the Dao and Principle; there is no further distinction between them. The 'all-embracing force' is my own force, which 'being nourished without sustaining injury, filled up Heaven and Earth'" (Fung, 1953, p. 510). Hence the human being can comprehend the Principle as what is above the shape, so long as he is willing to cultivate himself with sincerity and earnestness, and to practice the Dao in everyday life. For Cheng Hao and many other Neo-Confucians the human body-mind embraces all principles and all principles are complete in its Oneness. Such a view is consistent with what we have discussed earlier.

On the other hand, however, we must admit that some Neo-Confucians do make a apparent distinction between what is above the shape and what is within the shape. Cheng Yi, for instance, rejects any corporeal aspects of the Principle or the Dao, maintaining that the Dao only belongs to what is above shapes, and what is above shapes remains hidden. For Cheng Yi, the physical things (i.e., *yin* and *yang*) in the world are functional and fluctuating, whereas the Dao is constant and eternal. Therefore, the practice of *xiu shen*, according to Cheng Yi, is characterized by an "exhaustive study" of is the true knowledge, i.e., the Principle. A similar view can also be observed in Zhang Zai's critique of the "petty knowledge of hearing and seeing" vis-à-vis "good knowledge" (*liang zhi*), as well as in Zhu Xi's critique of the Buddhist views on the physical ability of seeing, hearing, speaking, thinking, and acting.[13]

In fact, Zhu Xi's argument on the body as within the shape is quite ambivalent. On the one hand, he argues that Nature, or what he calls the "Principle of Nature" or "Principle," is metaphysical, abstract and in the state of quiescence and unity, hence it cannot be physically perceived. On the other hand however, Zhu posits that the physical nature is no different from the nature of Heaven and Earth, and that the nature of Heaven and Earth runs through physical nature. Thus Zhu also argues that the process of *xiu shen* involves not only the Principle but also the manifestation of the Principle in each concrete thing in the world. Though Zhu holds a hierarchical view of the mind-body relationship when he regards the mind as the master of the body, the body as a vehicle to the external world has never been denied in Zhu's argument. This point can be seen in Zhu's rejection of the "yoke-carriage" metaphor when he speaks of the internal relation between "my mind" and "my body." Zhu argues that the relation of body and mind is like the body using the arm and the arm using fingers without making a distinction of which functions first. For Zhu, the ears, eyes, mouth, nose, and four limbs are parts of body, but only through the human mind can one see, hear, speak, or act. Meanwhile, the human mind depends on the functions of the sensory organs, the mind feels, wills, and functions through the body. One could argue that Zhu Xi's ambiguity mainly lies in his ambivalence toward the relationship between the mind and the Principle. When he equates the mind with an abstract notion of the Principle, he holds a mind/body duality, and looks for the Principle behind the body-shape. But when he separates the mind from an abstract notion of the Principle, he speaks of the interwovenness between the mind and the body, and emphasizes the possibility of transformation.

When asked how should we know the Principle, both Cheng Yi and Zhu Xi are back to what they call "investigation of things" through the "extension of knowledge." Here we recognize a movement from Principle to external things, and from things to Principle. This movement suggests a kind of cyclic activity, like an Odyssian journey where one leaves home for the purpose of coming back home. For Neo-Confucians the Principle is still a form of "presence." As Robert Magliola once indicated, "Confucianism, in effect the most 'worldly' of all major philosophico-religions, has also the most elaborate 'theory of presence' with each and every member of society related to each other and to Heaven, the Supreme Reality, in one glorious architectonic" (Magliola, 1984, p.90). In Neo-Confucian typology of the "above" and the "below", we see the polarity between the Principle and the structure of external things in the following diagram:

		Dao		
Heaven-Principle	←	**Xin**[a]	→	humanity-things
above the shape		↔		within the shape
Principle(*li*[a])		↔		life-energy(*qi*)
substance (*ti*)		↔		instrument (*qi*)
universal		↔		concrete
nature		↔		yin-yang
quiescence (*jing*[b])		↔		movement (*dong*)
unity		↔		multiplicity
mind		↔		**body**

In this Neo-Confucian distinction between "above" and "within" we find an attitude congenial to the reification and characteristic of totality of what we may call the "metaphysics of presence." And also, we can find within the vocabulary of Neo-Confucianism terms that bear binary structures such as reason-emotion (*xing-qing*), body-mind (*xin-xing*), and Principle-ether (*li-qi*). However, it is misleading if we simply say Neo-Confucianism is a form of Platonic metaphysics, because many Neo-Confucians do maintain the dialectical relationship between the Principle and myriad things, between substance and function. Accordingly, the Principle is seen as both the primordial source and that which embraces all principles. In other words, "it is one but the manifestations are many" (*li yi fen shu*, Smith, al. 1990, p.174). This notion of many-in-one and one-in-many is also a key concept in Chinese Buddhism, especially in Hua Yan School, which has had a strong impact on the Neo-Confucian idea of the relationship between the Principle of Heaven-Earth and myriad things. If for Cheng Yi and Zhu Xi, the One is still a name for the Principle world, the later Neo-Confucian thinkers after the *li xue* of Zhu Xi, such as Wang Fu-zhi (1619-93) and Yan Yuan (1635-1704), have replaced the One (the *Dao* or the Principle) with things in the world since they are more attracted to those that are "within the shape" in everyday affairs than were their predecessors. In his exposition of the relationship between the "above" and the "below" Wang Fu-zhi argues, "'Above' and 'below' are terms between which there is no fixed line of demarcation. 'What is above shapes' does not refer to an absence of shapes, but to the fact that there must be shapes before there can be 'What is above shapes.' If we probe into Heaven or Earth, man or other creatures, never, from antiquity until today, throughout all the myriad evolution, has ever been an 'absence of shapes above'" (Fung, 1963, p. 643). In his modality of cosmology, Yan Yuan contends, "The two forces (*yin* and *yang*) and fours powers conform, oppose, combine, and penetrate one another; they mingle, interweave, influence, warm, transform, change, arouse, stimulate, condense, disperse, roll, and unroll" (Fung, 63, p. 638). Here again, we see a "fusion" of "above" and "within," of the Principle (*li*) and things in the world (*xing*), although this fusion is articulated with a consciousness of difference. Therefore, the Neo-Confucian conception of "what is above the shape" is not simply a Platonic presence or a Cartesian *ego cogito* nor a merely corporeal body within the physical shape in a contemporary sense.

VI. CONCLUSION

As we have observed, *xiu shen* involves the physical well-being, ethical principles, and philosophical considerations. The Neo-Confucian discourse of *xiu shen* basically indicates a continuity of the Confucian understanding of the mind-body relationship in the light of human sensing, willing, knowing, and performing. The Neo-Confucian emphasis on *xiu shen* as a personal inner experience reinforces both the spiritual and pragmatic traditions in classical Confucianism.

Certain features in the Neo-Confucian concept of the body and its ethical sensibilities can be summarized as follows:

1. The human body is part of the onto-cosmological body that comprises the totality of Heaven-Earth, humanity, and myriad things;
2. The harmonious fit of one's body and mind is the goal of self-cultivation;
3. The temporality of the body requires the emphasis on every moment of existence and the fulfillment of one's duties in this life;
4. Experiencing and performing are the embodiment of the Dao;
5. The unity of the noumenal One to the phenomenal many is expressed in both the pattern of the Dao/Principle and its multiplicity implied in ethical principles in everyday life.

We should not be surprised if we find certain inconsistencies in Neo-Confucian thought because its immediate concerns and primary interests are ethical rather than logical. Though Neo-Confucians are attracted to such metaphysical inquiries as the Principle, the Supreme Ultimate, and internal cultivation, their fundamental concerns have never trespassed the realm of humanity. As Etienne Balazs, a French sinologist, once pointed out, "even when it [Confucianism] attempts to detach itself from the temporary world and arrives at some form of pure, transcendental metaphysics there can be no hope of understanding it without realizing its point of departure to which sooner or later it returns" (Tu, 1979, p.71).

Department of Religion
Temple University
Philadelphia, USA

NOTES

[1] Confucianism refers to the teachings of Confucius (551-479 B.C.) and (Mencius 372-289 B.C.). Neo-confucianism is also called the "learning of the Dao" (*dao xue*), or the "learning of the Principle" (*li xue*). The name itself suggests the linkage between the teachings of Confucius and Mencius and Neo-Confucian philosophy. In fact, Confucianism and Neo-Confucianism are virtually synonymous with traditional Chinese culture.

[2] Like Shao Yong, Wang Yang-ming claims, "The great man regards Heaven and Earth and myriad things as one body. He regards the world as one family and the country as one person." See Tu, 1979, p.80.

[3] The term "body-person," or a "personal body," is a term coined by the Japanese philosopher, Ichikawa Hiroshi. See Shigenori Nagatomo, 1993, pp. 321-346.

[4] For detailed a discussion, see Ann D. Birdwhistell's insightful interpretation in his 1989.

[5] See *Li Xue Zhuan Yao* by Jiang Bo Qian. Zheng zhong shu ju, Cheng Chung Book Company, 1982. pp.48-49.

[6] Shao Yong's theories of eternal recurrence and quieting sitting are inspired by Buddhism. In fact, many ideas in Neo-Confucianism are influenced by Buddhism although some Neo-Confucian thinkers, especially those of the *lixue*, are quite reluctant to admit their connections to the Buddhist tradition.

[7] The "mirror" is a symbol employed by Daoists, Chan Buddhists, and Neo-Confucianists. See Chan, 1973, p.207.

[8] The *lixue* is also called a "rationalistic school" because of its strong critique of emotions. See Fung, 1973, Vol. II, p.500.

[9] The term *zhong yong* means literally "centrality and universality." It has been translated as the Mean, mean-in-action, moderation, the middle way, etc. The Doctrine of Means is an important philosophical work for Neo-Confucian thinkers. According to Zhu Xi, *zhong yong* indicates a position that should be neither one-sided nor extreme.

[10] In many aspects, the understanding of body and death in Daoist philosophy is similar to that of Neo-Confucians. In the *Zhuang Zi*, for instance, we read: "The universe carries us in our bodies, toils us through our life, gives us repose with our old age, and rests us in our death. That which makes our life a good makes our death a good also." See Fung Yu-lan's *Chuang-Tzu*, p.116.

[11] Later Neo-Confucian philosophers in the Qing Dynasty such as Wang Fu-zhi and Yan Fu prefer to use the terms "substance" (*ti*[a]) and function (*yong*) to deploy the basic notion that Principle is one but also expressed by many. While *ti*[a] signifies the intrinsic qualities of a thing, *yong* refers its extrinsic manifestations. In the realm of ethics, *ti*[a] *i* indicates the inner value of humanity (*ren*), whereas *yong* stands for its functions and practice (*yi*[a]).

REFERENCES

Birdwhistell, A. D.: 1989, *Transition to Neo-Confucianism – Shao Yung on Knowledge and Symbol of Reality*, Stanford University Press, Stanford.

Chan, W.: 1973, *A Source Book in Chinese Philosophy*, Princeton University Press, Princeton.

Er Cheng Ji (The Collection of Essays by the Cheng Brothers): 1981, Zhong Hua Shu Ju, Beijing.

Fung, Y.: 1973, *A History of Chinese Philosophy*, Vol. I & II, Princeton University Press, Princeton.

Fung, Y.: 1994, *Chuang-Tzu: A New Selected Translation with an Exposition of the Philosophy of Kuo Hsiang,* Foreign Languages Press, Beijing.

Jiang, B.Q.: 1982, *Li Xue Zhuan Yao*, Zheng Zhong Shu Ju, Taipei.

Legge, J.: 1970, *The Works of Mencius*, Dover Publications, Inc., New York.

Magliola, R.: 1984, *Derrida on the Mend*, Purdue University Press, Indianapolis.

Nagatoma, S.: 1993, 'Two contemporary Japanese views of the Body: Ichikawa Hiroshi and Yuasa Yasuo,' in Thomas P. Kasulis, Roger T. Ames, and Wimal Dissanayake (eds.), *Self as Body in Asian Theory and Practice*, State University of New York Press, Albany, pp.321-346.

Smith, *et al.*: Sung Dynasty Uses of the I Ching, Princeton University of Press, Princeton.

Tu, W.: 1973, *Humanity and Self-Cultivation: Essays in Confucian Thought,* Asian Humanities Press, Berkeley.

Tu, W.: 1992, 'A Confucian perspective on embodiment,' in Drew Leder (ed.), *The Body in Medical Thought and Practice*, Kluwer Academic Publishers, Dordrecht, pp. 87-100.

PART II

SUICIDE, EUTHANASIA AND MEDICAL FUTILITY

PING-CHEUNG LO

CONFUCIAN VIEWS ON SUICIDE AND THEIR IMPLICATIONS FOR EUTHANASIA

I. INTRODUCTION

On June 2, 1927, a famous professor of Qing Hua University, Wang Guowei, drowned himself in a lake of the former imperial garden in Beijing. His suicide proved extremely controversial and evoked much discussion.[1] His colleague and famous intellectual, Liang Qichao, wrote several eulogies in his honor. In one of these eulogies Liang reminded his colleagues and students, who lived in a culture which had just entered the modern age and was under heavy western influence, not to use western perspectives to evaluate Wang's suicide. Europeans, Liang asserted, used to regard suicide as an act of cowardice, and Christianity made it a sin. In ancient China, however, notwithstanding some petty suicides committed by common people, many eminent figures used suicide to express their counter-cultural aspirations. These were praiseworthy suicides, Liang concluded, and should by no means be rashly reproached by alien European values (Liang, 1927, p.75).

I think the main idea of Liang's observation is largely correct. Traditional Chinese ethical perspectives on suicide differ significantly from their western counterparts, with the possible exception of Stoic Rome. Just as western views on suicide have been strongly influenced by the ancient Greek philosophers (Pythagoras, Plato, and Aristotle) and Augustinian Christianity, the Chinese views, especially among the educated people, have been largely influenced by classical Confucianism. Hence, in this paper I shall try to provide a comprehensive, analytical survey of the Confucian ethics of suicide.

Some distinction among suicides is necessary for an in-depth ethical analysis because not all suicides are morally equivalent.[2] Emile Durkheim's three-fold typology of egoistic, altruistic, and anomic suicide is well known (Durkheim, 1951, Book Two, esp. pp.209, 221, 258). This paper, however, is not a study on the social causes of suicide, and hence cannot adopt his classification. For my purpose in this paper, I shall classify suicide into two types, viz., suicides committed for self-regarding reasons (to be abbreviated as self-regarding suicides below) and suicides

Ruiping Fan (ed.), Confucian Bioethics, 69–101.

committed for other-regarding reasons (to be abbreviated as other-regarding suicides below).[3] Two explanations are in order before I proceed forward. First, in other-regarding suicides the reference of "other" can range from one individual to the entire country. Second, the meaning of "other-regarding suicide" is wider than that of "altruistic suicide" because altruism is consequence-oriented, viz., the promotion of others' interests. An other-regarding suicide can be consequence-oriented, but not necessarily so; it can be a suicide for the sake of manifesting one's total dedication to another person or persons.

I submit that in the premodern West (Europe before the Enlightenment), the focus of the relevant moral debate was largely, though not exclusively, on self-regarding suicide. And this debate led to a predominantly negative moral judgement of this type of suicide.[4] To approve of suicide morally had the burden of apology, and famous defenses of the right to suicide were largely concerned with self-regarding suicide (e.g., Seneca, 1920; Hume, 1965). The major moral issue was, accordingly, "Is it morally permissible to commit suicide, especially suicide for one's own sake?"

In premodern China (China before 1911), most self-regarding suicides were also generally regarded as wrong, but unlike the premodern West there was virtually no defense of an individual's right to suicide. There were some discussions on a special kind of self-regarding suicide, i.e., suicide for the sake of preserving dignity. However, most of the moral disputations were focused on other-regarding suicide, which was not uncommon in ancient China. From ancient Chinese perspectives, such a suicide was never deemed wrong and needed no apology; those who thought otherwise, however, had the burden of proof. Famous suicide-related apologies were defenses for the moral right not to commit suicide, and they were largely pleas for exemption from the moral demand of committing an other-regarding suicide in certain situations (e.g., the defense for Guan Zhong, see Section VI below). The major moral issue was, accordingly, "Is it morally permissible not to commit suicide in certain situations, especially suicide for others' sake?"[5]

II. THE VIEW OF EARLY CONFUCIANISM AND ITS INFLUENCE

A distinct emphasis in early Confucian ethics is that biological life is not of the highest value. Confucius (c.551-479 BC) says:

> For gentlemen of purpose and men of *ren* (benevolence or supreme virtue) while it is inconceivable that they should seek to stay alive at the expense of *ren*, it may happen that they have to accept death in order to have *ren* accomplished (*Analects*, 15:9, Lau (trans.) modified).

Likewise, Mencius (c.372 – 289 BC), the second most famous Confucian after Confucius,[6] also explains this in this famous passage on fish and bear's palm.

> Fish is what I want; bear's palm is also what I want. If I cannot have both, I would rather take bear's palm than fish. Life is what I want; *yi*[a] (justice or dutifulness) is also what I want. If I cannot have both, I would rather take *yi*[a] than life. On the one hand, though life is what I want, there is something I want more than life. That is why I do not cling to life at all cost. On the other hand, though death is what I loathe, there is something I loathe more than death. That is why there are dangers I do not avoid... Yet there are ways of remaining alive and ways of avoiding death to which a man will not resort. In other words, there are things a man wants more than life and there are also things he loathes more than death. This is an attitude not confined to the moral person but common to all persons. The moral person simply never loses it (*Mencius*, 6A:10, Lau (trans.) modified).

These two discourses together became the *locus classicus* of the classical Confucian view on the value of human life, and have been tremendously influential down through the ages. According to this classical view, the preservation of our biological life is a good, but not the supreme good; death is an evil, but not the supreme evil. Since the cardinal moral values of *ren* and *yi*[a] (benevolence and justice) are the supreme good, it is morally wrong for one to preserve one's own life at the expense of ignoring *ren* and *yi*[a]. Rather, one should sacrifice one's life, either passively or actively, in order to uphold *ren* and *yi*[a] (as to be explained later, in the form of upholding some important interpersonal values such as filial piety or loyalty to the emperor, discharging the related moral duties, and cultivating one's virtue or moral character). The failure to follow *ren* and *yi*[a] is ethically worse than death. Hence suicide is morally permissible, and even praiseworthy, if it is committed for the sake of *ren* and *yi*[a]. In some circumstances, furthermore, committing suicide is more than supererogatory; it is even obligatory. There is a doctrine of the sanctity of moral values, but not a doctrine of the sanctity of human life. Sheer living has no intrinsic moral value; living as a virtuous person does. There is no unconditional duty to preserve and continue life, but there is an unconditional duty to uphold *ren* and *yi*[a]. In English the phrase "matters of life and death" carries the connotation of urgency and utmost importance. According to classical Confucian ethical thought, however, though "matters of life and death" are by no means trivial, they are not of paramount importance; "matters of *ren* and *yi*[a]" are. For the sake of convenience, I shall summarize this classical Confucian view as Confucian Thesis I:

> **Confucian Thesis I: One should give up one's life if necessary, either passively or actively, for the sake of upholding the cardinal moral values of *ren* and *yi*[a].**

The meaning of this thesis can be further elaborated by studying the thought of Wen Tianxiang (1236-1282). When the Mongols invaded China in the thirteenth century and were about to conquer the entire land of the Southern Song Dynasty, many generals preferred suicide to surrender. The most famous example among them was Wen Tianxiang, who kept a suicide note in his pocket at all times during his capture, which began: "Confucius says that one should fulfil *ren*, and Mencius says that one should adhere to *yi*[a]." In a short poem he also wrote, "Who does not have to die (in one way or another) since time immemorial? (The preferable way of dying is) that my heart of pure loyalty may leave a page in the annals" (*Songshi*, Biography 177). This saying became immensely popular in subsequent times, known by most Chinese, and remains so even today.

Wen's point is that since everyone has to die, one should not try to avoid or delay death by all means. Longevity is not good in itself; rather, what is desirable in itself is a life of *ren* and *yi*[a], a life that will be remembered, respected, and honored in history. If, in some circumstances, staying alive would be contrary to the requirements of *ren* and *yi*[a] (in Wen's case, to stay alive would mean to surrender to the Mongols, which is contrary to the virtue of loyalty to the Southern Song Dynasty), it would follow that one could lead a life of *ren* and *yi*[a] only by committing suicide. If one has to die one way or another, one should die in such a way that can render one's life meaningful or honorable. In other words, though death is the termination of life, dying is still a part of life. "How one dies" is part of "how one lives." Hence, dying should serve life. To take charge of one's life implies to take charge of one's dying. To secure a noble and honorable life implies that one should secure a noble and honorable death. To live meaningfully implies to manage the time and circumstances of one's death in such a way that one can also die meaningfully. To live out one's life to its natural limit is not in itself desirable. What matters is not life's quantity (its length), but its quality, to be defined morally with reference to *ren* and *yi*[a]. In order to secure a high quality of life, in some circumstances one has to be prepared to die, lest what is going to transpire in one's prolonged life will decrease the quality of life (i.e., to violate *ren* and *yi*[a]).

A large number of sayings in premodern China reflect this classical Confucian view. For example:

> "A *junzi* (person of noble character) does not drag on his life indiscriminately, nor does he terminate his life indiscriminately";
>
> "To go to die while one should remain alive is to treat one's life too lightly; to live on while one should die is to treat one's life too weightily" (Li Bai);
>
> "A crooked life is not enjoyable; an upright death is not lamentable" (Han Yu);
>
> "A death of *yi*[a] (righteousness) is not fearsome, and a fortunate and surprising survival is not honorable" (Han Yu);
>
> "A death of *yi*[a] (righteousness) is to be preferred to a dragged out, ignoble existence" (Ouyang Xiu);
>
> "Rather to lose one's life while preserving one's moral integrity than to preserve one's life while losing one's moral integrity" (from an obscure ancient novel);
>
> "A brave general does not barely manage to survive out of fear for death; a heroic warrior does not seek staying alive at the expense of moral integrity" (Guan Yu, in *The Romance of the Three Kingdoms*).[7]

All these sayings express the view that one should aspire to live a moral life, and that there are states of biological existence that should be avoided even by seeking death. The quality of moral life is much more important than the length of biological life.

In short, this Confucian teaching of "to die to achieve *ren* (*shashen chengren*)" and "to lay down one's life for a cause of *yi*[a] (*shesheng quyi*)" not only has inspired countless Chinese to risk their lives for noble causes, but also has motivated many Chinese to commit suicide for noble causes. When people thus committed suicide, they were not condemned; rather, they were praised for their aspiration and dedication to *ren* and *yi*[a]. This is the case even as late as in the early twentieth century. The suicides of two intellectuals, Liang Juchuan (d. 1918) and Wang Guowei (d. 1927), were eulogized for exhibiting *ren* and *yi*[a] (in Luo, 1995, pp.54, 63). Even Chinese Buddhist biographers repeatedly quoted the Confucian *locus classicus* to justify acts of self-immolation of Buddhist monks in ancient China (Jan, 1964, p.260; see also Section IV below).

There is one other articulated view on life and death in ancient China that has also been immensely influential. Though the one who expressed this view, Sima Qian (c.145–190 BC), the Grand Historian, has not been known as a Confucian, his admiration of Confucius is obvious in his *Shiji* (*Records of the Historian*). In his famous *Bao Ren An Shu* (*Letter to Ren An*), which is a confession of a soul tormented by the thought of suicide, he makes the memorable statement, "Everyone has to die sooner or later. Whether the death is weightier than Mount Tai or lighter than swan's down depends on its circumstance" (Sima, trans. mine).[8] In other words, everyone has to die; that is the same for all. The value of their death, however, is not the same. Some deaths are good, while some are of no value, or even bad. The degree of value depends on the circumstances of the death. If committing suicide in a particular circumstance would be of significant value (i.e., weightier than Mount Tai), one should do it. One should not commit a suicide that would have little significance (i.e., lighter than swan's down). In other words, according to Sima Qian, death is not a bare biological event, at least as far as human beings are concerned. The time and circumstances of one's death have ethical significance. The moral issue is not whether one can commit suicide for there is no strict prohibition against it, i.e., suicide is not intrinsically wrong. Rather, the issue is for what kind of reason (trivial or substantial) the suicide is committed, and what kind of impact it will produce. In strictly Confucian terms, suicide committed for the sake of *ren* and *yi*[a] would be a death weightier than Mount Tai. (More on Sima Qian will be discussed in the next section.)

The obvious and important questions, then, are: What is the content of *ren* and *yi*[a]? How are these cardinal values embodied in concrete actions? What suicides can be counted as suicide for the sake of *ren* and *yi*[a]?

Both *ren* and *yi*[a] have a narrow and a wide sense. In the narrow sense, as the first two of the four cardinal virtues, *ren* means benevolence and *yi*[a] means justice. In the wide sense however, both words, especially when they are used together, can mean supreme virtue or morality (cf. Nivison, 1987, pp.566-567). In the context of "to die to achieve *ren*" and "to lay down one's life for a cause of *yi*[a]," *ren* and *yi*[a] were usually understood in the wide sense. One should note however that since the Han Dynasty morality, or *ren* and *yi*[a], has been conceived of manifesting itself in particular human relationships, rather than in a universal and general way. In other words, *ren* and *yi*[a] were understood not through universal love or duty to society in general, but through interpersonal commitments such as loyalty (in the emperor-subject relationship), filial piety (parents-children relationship), chastity (husband-wife relationship), and faithfulness (friendship). In other words, *ren* and *yi*[a] are concerned with other-regarding morality and are mediated through concrete familial, social, and political relationships.

Accordingly, all other-regarding suicides, when the "other" is a particular object rather than human beings in general, are deemed suicides for the sake of *ren* and *yi*[a]. They were admired, praised, and honored. Among the more important instances are:

- suicide for the sake of the country (dynasty) and/or emperor;
- suicide for the sake of the husband, who just passed away;
- suicide for the sake of the master;
- suicide for the sake of the benefactor, as a token of gratitude;
- suicide for the sake of a friend, especially those with whom one has entered into a pact of brotherhood;
- suicide for the sake of keeping a secret for somebody;
- suicide for the sake of saving other lives;
- suicide for the sake of avenging one's parents, husband, or master.[9]

It should be noted that these other-regarding suicides can be divided into two ethical types, viz., consequence-based and non-consequence-based. An example of a consequence-based other-regarding (or altruistic) suicide is that of a public official who commits suicide in order to awaken and admonish a self-indulgent emperor. An example of a non-consequence-based other-regarding suicide is that of a public official who follows the emperor even unto death or that of a widow who chooses to die with her deceased husband. Total dedication, devotion, and commitment to the other person leads one to live and die with the other person, even though such a suicide is not intended to benefit anyone. Either way, such other-regarding suicides were expressions of utterly other-centered commitments, even at the expense of one's life. Such self-sacrificial acts, understandably, were highly praiseworthy.[10]

All these suicides were considered suicides of *ren* and *yi*[a]. Furthermore, terminologically speaking these acts were not called "suicide" in the pejorative sense of "self-destruction" or "self-slaughter."[11] A different set of phrases, usually a combination of another word with *xun* (sacrifice)[12] or *jie* (moral integrity)[13] with a connotation of praiseworthiness, was used instead. Tang Jungyi, a contemporary neo-Confucian, compares such death of *xun* and *jie* to the death of martyrs in early Christianity. Just as the Christian martyrs were prepared to endure anything, even death, for the sake of upholding their faith, the Confucian men and women of integrity (*qijie*) were also prepared to endure anything, even death, for the sake of upholding *ren* and *yi*[a]. Tang Jungyi therefore adds that the religiosity (absolute devotion, unconditional dedication, ultimate commitment) of these men and women could not be denied (Tang, 1974, p.144).

In addition to the detailed examples to be discussed in the other sections of this article, I shall cite two more illustrations of praiseworthy suicides here. First, during the Yuan Dynasty (1279 – 1368) there was a famous opera entitled *The Orphan of Zhao* (for a complete English translation, see Chi, 1972), whose plot was based on accounts in historical books. (In other words, many Chinese before the Yuan Dynasty knew the story as well.) The Jesuit missionaries in China later brought this opera to Europe, where it was immensely popular and translated into English, French, and German (cf. Hsia, 1988). Voltaire not only rendered it into a drama, with the name changed to *L'Orphelin de la Chine*, but also staged it successfully in Paris. The German philosopher Arthur Schopenhauer also mentioned this play with admiration in his essay "On Suicide" (Schopenhauer, 1962, p.99). The story is about a family of nobility being persecuted by its political enemies, and everybody in the family is killed except a baby. Some friends of the family try their best to save the life of this orphan. In the process, all the noble characters who take part in this saving effort commit suicide, mostly for the reason of ensuring the rescue effort will be successful. These are all altruistic suicides. At the end of the story, after the rescued orphan has grown up and avenged his father, the architect of the rescue and vengeance, Cheng Ying, also commits suicide for the reason of going to the underworld to tell all those who committed suicide for the orphan that their deaths were worthwhile.

The second illustration is taken from recent Chinese history. On the eve, and during the birth, of modern China there were a number of suicides that produced an impact on and evoked much discussion among many intellectuals. Liang Juchuan (d.1918), Wang Guowei (d.1927), both self-confessed Confucians, were two of them.[14] However, I want to discuss the suicide of Chen Tianhua (1875-1905) because his life and detailed suicide note provide a clear exposition of Confucian Thesis I.

Toward the end of the Qing Dynasty some Chinese, both inside China and overseas, plotted to start a revolution to overthrow the Manchu rule. Chen Tianhua was an enthusiastic supporter of this movement and composed many tracts and pamphlets to spread the idea of the necessity of revolution. During his brief study in Japan he was convinced that Chinese students studying overseas were the only hope to save the motherland from the threat of imperialism. However, he was subsequently overwhelmed by the realization that a number of Chinese students in Japan were involved in scandalous behavior, to the extent that they were widely criticized by Japanese newspapers. Chen was deeply grieved, and wrote a long suicide note ("*jueming ci*") before drowning himself in the sea.

In this suicide note he explains clearly his reasons for this shocking act. (1) He considered himself a man of mediocre ability; he would be unable to accomplish great deeds for the country even if he was to live a long life. (2) His future life could have two possible directions: either to continue to write as a doomsday prophet, warning people that if they would not pull themselves together China would perish, or to die if the right circumstance was to arise. (3) He seriously considered the latter possibility because, even though his speeches and writings were quite influential among the Chinese students in Japan, he did not want to be one of those who is stirring in words but seldom matches these words with practice. (4) He was fully determined to spend his life striving to save the country, and he needed to deliberate carefully which course of life would be most effective to achieve this goal. (5) Given the current situation in Japan, viz., the Chinese students were lax in their conduct and did not show intense zeal for the country's cause, Chen Tianhua concluded that he could contribute more to the cause of saving China by committing an admonishing suicide. By choosing to die Chen wanted to admonish the eight thousand Chinese students in Japan to pull themselves together, to dedicate their lives to the country's cause. (6) Chen reiterated that he was not a man of outstanding ability, and he therefore could accomplish more by dying than by living a long life. Those who had better talents should not follow his example (T. Chen, 1982, pp.234-236).[15]

Though there is no hard evidence that Chen Tianhua was a self-confessed Confucian, I think his duty-bound death was a clear exposition of Confucian Thesis I. *Ren* and *yi*[a] in his particular case were concretized as saving the country from perishing. He was so dedicated to this cause that he was willing both to live and to die for it. In other words, the same patriotic life goal can call him to stay alive or call him to die. His decision to commit suicide did not reflect a change of his values, but depended entirely on the empirical circumstances. The time and circumstances of one's death was considered a part of life and important to its goal. Biological life is not intrinsically valuable or an end in itself, but only instrumentally valuable and a means to serve a moral cause. Hence suicide is not intrinsically wrong. If terminating biological life can better serve this moral cause, one should die rather than stay alive. Chen's suicide was not an act of self-destruction; rather, it was an act of moral construction.

There were, of course, many suicides in premodern China that were not deemed suicides for the sake of *ren* and *yi*[a]. Most self-regarding suicides were not so evaluated, e.g., suicide as a result of being tired of life, suicide as a solution to one's financial or marital troubles and failures, suicide as a solution to chronic depression, suicide as an expiation of one's wrongdoing, suicide out of a fear of punishment or public mockery, etc. They were pitied and deplored. These suicides were evaluated as "self-destruction" or "self-slaughter," and many of them were deemed wrong primarily because of another important Confucian value, viz., *xiao* or filial piety. Committing suicide was deemed contrary to filial piety not because of the trivial reason that it would cause grief to one's parents.[16] For one thing, Confucian filial piety requires that sons and daughters should attend to the parents' daily needs all life long. Terminating one's life would render one unable to fulfil this important filial duty. For another, Confucian literature on filial piety argues that children are permanently indebted to parents because children do not exist on their own, but owe their existence to the parents. If one is not the author of one's biological life, how can one have the autonomy to dispose it as one wishes? Suicide is then understood as usurping the authority of parents.[17] In short, unless filial piety is outweighed by another moral value such as *ren* or *yi*[a], the former is usually a moral reason strong enough to forbid suicide.

III. DEATH WITH DIGNITY

Most self-regarding suicides, as explained in the last paragraph, were generally regarded as morally wrong in ancient China, and nobody felt the need to discuss them further. There was one kind of self-regarding suicide, however, that did evoke some discussion, and it can be conveniently called with a modern idiom, "death with dignity." In the Former Han Dynasty (206 BC – 8 AD) Confucianism was elevated to the role of the established ideology of the empire. The Confucian who was instrumental in making this happen was Dong Zhongshu (c.179 – c.104 BC). Though modern Chinese philosophers often consider him of minor philosophical significance, historically he was of utmost importance. The imperial policy of establishing the supremacy of Confucianism to the exclusion of other schools of thought, which was advocated by Dong, was adopted in 136 BC and was continued until 1905 AD (i.e., this policy of establishing Confucianism was in effect for a little more than two thousand years).[18]

One should note that it was not the original classical Confucianism that was honored in the Han Dynasty, but Dong's creative synthesis of various streams of Confucianism together with other schools of thought. Dong's masterpiece was entitled *Chunqiu Fanlu* (*Exuberant Dew of the Spring and Autumn*), which was an exposition of the thought of the *Spring and Autumn Annals*, the author of which was said to be Confucius. Dong regarded the *Spring and Autumn Annals* as the primary text within the Confucian canon. In his *Exuberant Dew of the Spring and Autumn* Dong eloquently elaborates on a variation of Confucian Thesis I, which was shared by other Confucian writings around the same time in the early Han Dynasty.

In Chapter 8 ("Zhulin") of the *Exuberant Dew of the Spring and Autumn* Dong discusses a certain king and his adviser who lived several hundred years before his time. King Qing of Qi was in a battle with his enemies, and lost. The enemies surrounded his armies and it became highly likely that he would be captured and killed. His adviser Choufu happened to look like King Qing and therefore offered to exchange clothing with him so that he could escape unnoticed. The strategy succeeded. King Qing escaped back to his kingdom in civilian clothing while Choufu was mistaken to be the king, and was captured and killed.

Dong Zhongshu, rather than praising Choufu's ingenuity, dedication, and sacrifice, condemned his action. To get a king to dress as an ordinary citizen and escape surreptitiously, according to Dong, was to subject a dignitary to an undignified treatment. Such humiliation should not be tolerated, even if it could save life. This is because, Dong argues, "to survive through accepting a great humiliation is joyless, thus wise people refrain from doing it ... A person who has a sense of shame does not live in dishonor." He also quotes from other Confucian writings of the early Han Dynasty, implying that his ethics of suicide was derived from the Confucian canon. "If a dishonor is avoidable, avoid it; if it is unavoidable, *junzi* (a man of noble character) sees death as his destiny (i.e., embraces death with courage) ... A *ru* (Confucian) prefers death to humiliation."[19]

Dong therefore argues that the morally right thing for Choufu to do then would be to tell King Qing, "To bear humiliation and yet refuse to commit suicide is shameless. I shall therefore commit suicide with you." At that moment, for both of them, death would be better than staying alive as "a *junzi* (man of noble character) should prefer dying in honor to surviving in dishonor."

In short, according to Confucianism of the early Han Dynasty, biological life is valuable, but there are self-regarding states of affairs more valuable than biological life, viz., a life with honor and dignity. Death is undesirable, but there are self-regarding states of affairs more undesirable than death, viz., to suffer disgrace, dishonor, and humiliation in life; such a life is not worth living because it is such an affront to one's dignity. One should choose death for the sake of preventing one's dignity from being violated, and it is honorable, and even obligatory, to make such a choice. This view is a variation and elaboration of Confucian Thesis I, with the focus shifted from other-regarding concerns to self-regarding concerns. For the sake of convenience, I shall call it Confucian Thesis II:

Confucian Thesis II: One should actively terminate one's life for the sake of avoiding humiliation or upholding one's dignity.

In the light of Confucian Thesis II, I suggest that we can use the phrase "death with dignity" to describe this view of committing suicide in order to preserve one's dignity, i.e., voluntary death as a means to maintain one's dignity. In other words, "death with dignity" in this sense is not to counter a death without dignity, but to preclude a life that is deprived of its usual dignity. Death is chosen for the sake of preventing one's dignity from being violated, and it is an honorable and moral duty to make such a choice.

Such "death with dignity" was quite common in ancient China and many examples could be found in the *Records of the Historian* (*Shiji*) by Sima Qian (145-90 BC), the greatest historian of ancient China and a younger contemporary of Dong Zhongshu. In the *Records of the Historian* many suicides were recorded, and often with approval. Among these suicides two types are particularly noteworthy for our purpose.

(I) One commits suicide when death is unavoidable in the near future.
a. One hears or predicts that one will be executed by the government, and so commits suicide.
b. One commits suicide after a military defeat (otherwise the defeated general will be killed by his conqueror).
c. One commits suicide after a failed *coup d'etat* attempt (which means that execution is merely waiting for the rebel).

What is common in all these three cases is that the fate of execution is considered a humiliation, a dishonor, and a disgrace. Hence it is better to kill oneself than to be killed by others. Committing suicide is therefore a means of preserving one's dignity.

(II) One commits suicide when there is no known threat to life.
a. A Confucian public official commits suicide in order to avoid the indignity of being tried in court, regardless of whether he is guilty or innocent.
b. A Confucian public official commits suicide in order to avoid the indignity of imprisonment.

These Confucian public officials firmly believe that to be tried in court and/or to be imprisoned, even if one is innocent, is a humiliation, a dishonor, and a disgrace. Hence it is better to kill oneself than to suffer such an undignified treatment. Committing suicide is therefore a means of preserving one's dignity. In short, both types of suicide can be considered a "death with dignity," and they suggest that Confucian Thesis II was widely accepted in Chinese antiquity.

It is noteworthy that Confucian Thesis II was even accepted by some contemporary intellectuals in China. During the Cultural Revolution

(1966-1976) many university professors and men and women of culture were publicly tortured, brutalized, and humiliated. Many of them committed suicide (e.g., Fu Lei, Lao She).[20] Some could not stand the physical and emotional suffering, while some simply would not accept this humiliation. A famous senior philosophy professor at Peking University told me in 1993 that he knew a colleague who used to take these purges calmly. One morning, however, he discovered an accusation poster posted on his front door that was written by his students. He was deeply hurt. He left a note stating that "a man of integrity prefers death to humiliation," then he committed suicide. This shows that Confucian Thesis II is also accepted by some contemporary Chinese.[21]

To commit suicide for the sake of preserving personal dignity, though self-regarding rather than other-regarding, was usually met with approval and even admiration. This mainstream view, however, was also countered by a dissenting view. For example, Sima Qian, who approved of many "death with dignity" suicides in his *Records of the Historian*, rejected this option when he himself was put in the predicament of receiving extremely undignified treatment, viz., he suffered castration in prison. This happened because Sima Qian once defended a general who surrendered to the "barbarians" after a military defeat. The subsequent development of events convinced the emperor that this general was a traitor and punished all those who once pleaded for him. Sima Qian was therefore imprisoned and castrated. Sima Qian considered this punishment a humiliation to the umpteenth degree, and understood that his peers expected him to commit suicide in order to avoid this undignified treatment. Sima Qian, however, after many struggles, refused to commit suicide not because he did not care about his dignity but because he decided to bear this unbearable indignity in order to complete his half-finished masterpiece, the *Shiji*. He understood very well that he had a duty to commit suicide, but he considered it more important to discharge his weightier duty of writing a grand historical book. His *A Letter to Ren An* can be read as the account of a tormented soul urging his contemporaries to excuse him for not committing suicide.

In short, in the midst of undignified treatment, death with dignity is not the only option. One can continue to live on with dignity by fulfilling one's vocation rather than just dragging on. There are many historical examples of this nature, as Sima Qian also noted in the *Letter*. Sima Qian's decision of not committing suicide has also been very influential in the subsequent development of Chinese thought. His view can be summarized as Antithesis II, in sharp contrast to Confucian Thesis II:

> **Antithesis II: When there is no threat to one's life, and when the calling in life is clear, one should live on to fulfill one's vocation in spite of personal tragedy and undignified treatment.**

IV. THE INFLUENCE OF DAOISM AND BUDDHISM

The Confucian ethics of suicide was, of course, not uncontested in ancient China. To understand the distinctiveness of the Confucian perspective, it is best to contrast it with the Daoist and Buddhist perspectives.

In Daoist philosophy, both Lao Zi and Zhuang Zi (c.369-286 BC) were critical of Confucian ethics. Zhuang Zi, in particular, criticized the Confucian view on suicide for the sake of *ren* and *yi*[a] in chapter eight ("*pianmu*," or "Webbed Toes") of his writings (Chuang Tzu, 1964). According to him, the morality of *ren* and *yi*[a] is external to human nature; it is artificial, not natural. To die for *ren* and *yi*[a], therefore, is to blight one's inborn nature, which is equivalent to dying for profit or reputation. One should, instead, preserve one's inborn nature; life is an end in itself, not a means to morality. (As I shall explain in the next section, Jia Yi used this Daoist argument to lament the suicide of Qu Yuan.)

The Daoist religion is even more adamant in its negative attitude to all kinds of suicide because its religious goal is to attain immortality of life. To end one's life intentionally shows one's ignorance of the supreme goal of life. In short, both Daoist philosophy and religion consider suicide intrinsically wrong (Zheng, 1994).

Buddhism, on the other hand, is not as unsympathetic to suicide as Daoism. This is because, according to the Buddhist worldview, death is only a transitional point in *samsara*, a transition between different forms of karmic existence. Hence the crucial issue is not when one dies, but how one dies – whether the manner of death will cause merits or demerits in the next form of life. Suicide is therefore to be evaluated according to its motive (De La Vallee Poussin, 1908).

Consequently, even though early Buddhism in India once strictly forbade suicide, altruistic suicide was highly praised in Indian Mahayana Buddhism when suicide was committed for the sake of almsgiving (e.g., to kill oneself to feed a hungry tigress, lest she will devour her baby cubs). Such almsgiving (*dana*) of the body was a manifestation of supreme compassion (*karuna*) to all living beings (Dayal, 1932, pp.178-188). In China, Mahayana Buddhist other-regarding suicide took on a new form, viz., to burn oneself to death as incense offered to the Buddha. Quite a number of eminent monks committed suicide this way, and found the inspiration from chapter twenty-two of the *Lotus Sutra*. (*Biographies of Eminent Monks*, *More Biographies of Eminent Monks*, and *Biographies of Eminent Monks in Song Dynasty* all contain a section of monks who died in this manner.)[22]

V. QU YUAN

Confucianism, of course, has not been monolithic. Even among the premodern Confucians, there were debates on the moral evaluation of certain suicides. Qu Yuan's suicide and Guan Zhong's refusal to commit suicide, in particular, engendered a long history of debate. I shall discuss the former in this section, and the latter in the next section.

The only national festival that commemorates a person in ancient and contemporary Chinese society is the Duan Wu Festival (Dragon Boat Festival). It commemorates the death of Qu Yuan (c.339-c.278 BC), who died by drowning himself in a river. The fact that he committed suicide not only did not lead to a stigmatization, but on the contrary it earned him great moral fame.[23]

According to tradition, two different reasons were advanced to explain his suicide. First, it was said that he hoped to use his sudden death to awaken the Chu king to realize that the country was in a crisis and to launch the badly needed political and military reforms. Second, it was said that he believed that his country was on the verge of conquest by the Qin kingdom, and therefore he wanted to die with his motherland. Either way, his suicide was other-regarding and a manifestation of his patriotism, and so it was deemed a suicide for the sake of *ren* and *yi*[a], an illustration of Confucian Thesis I. In the course of Chinese history, many people followed his example. For instance, Wang Guowei's drowning, mentioned at the beginning of this paper, was often regarded as an imitation of Qu Yuan (see Luo, 1995, pp.56-57, 63, 67, 70, 80).

Throughout Chinese history, though the mainstream opinion has been to praise Qu Yuan for his death, there has also been a stream of dissenting opinion that does not regard his suicide as praiseworthy. Two different kinds of reasons have been advanced. First, there is a Daoist argument which considers suicide, even other-regarding suicide, a harm to one's inborn nature; it violates the imperative to preserve one's life. For example, the Han writer Jia Yi (200-168 BC) expresses this opinion in his *Diao Qu Yuan Fu* (*Elegy to Qu Yuan*). A phoenix flies away when there is danger so as to preserve its life. A dragon, in treasuring its own worth, continues to stay in the bottom of a deep pool; it emerges and flies into the sky only if there is cloud in which to make its abode. Likewise, given his talents and noble character, Jia Yi argues that Qu Yuan should cherish and preserve his own life. Second, there is a cost-and-benefit argument which claims that had Qu Yuan stayed alive he could have made much more significant social, political, and cultural contribution, either to Chu or to another Chinese kingdom. To die for a fatuous king is not worthwhile. For example, the Han thinker Yang Xiong (53-18 BC) expresses this view in his *Fan Li Sao* (*Antithesis to The Lament*). He points out that though Confucius also had the aspiration to be a successful politician in his home country Lu, he did not cling to Lu. When circumstances did not allow him to carry out his ideal Confucius left Lu and traveled to the neighboring kingdoms to look for a ruler who would appreciate his ability. Qu Yuan should have done likewise, according to Yang Xiong. Even Zhu Xi, the greatest Confucian philosopher in medieval China, had misgivings on Qu Yuan's suicide. In his critical comments on Yang Xiong's *Fan Li Sao* (in his *Chuci jizhu, or Collected Commentaries on Chu Poetry*) Zhu nonetheless concedes that Qu Yuan's suicide was an act of excessive loyalty. Zhu concludes that although Qu Yuan's general integrity was perfect, there was a flaw in his particular deeds, viz., his excessive loyalty was contrary to the Confucian Doctrine of the Mean.

In other words, throughout Chinese history the death of Qu Yuan has been a topic of debate (see Huang, 1987). The majority opinion, under the influence of Confucian Thesis I, considers his other-regarding suicide praiseworthy. The minority opinion argues that either there is no duty to die in that situation, or that the duty is overridden by other duties; hence it pleads that such a suicide should not be elevated to a moral height. (This divergence of moral evaluation can be further ethically analyzed by making use of different schools of Confucianism, and of the "Three Forms of Immortality," to be explained below.)

VI. GUAN ZHONG

Another long debate centered on Guan Zhong (?-645 BC), a distinguished ancient statesman. Qi was a vassal state during the Spring and Autumn Period. There were two brothers who contended to succeed the throne of their father. Guan Zhong and Shao Hu assisted Prince Jiu, whereas Bao Shuya assisted Prince Xiaobai. The latter prince managed to succeed his father and came to be known as Duke Huan of Qi. He ordered the death of his brother, Prince Jiu, and Shao Hu followed his master to death by committing suicide. Guan Zhong not only did not commit suicide as Shao Hu did, but also switched allegiance to serve as the adviser to Duke Huan. He proved himself to be an exceptional statesman and helped Qi to dominate other vassal states and stabilize the Eastern Zhou Dynasty.

There is an interesting discussion of Guan Zhong in the *Analects* between Confucius and his two outstanding disciples, Zi Lu and Zi Gong.

> Zi Lu said, "When Duke Huan had Prince Jiu killed, Shaohu died for the Prince but Guan Zhong failed to do so." He added, "In that case, did he fall short of *ren*?" The Master said, "It was due to Guan Zhong that Duke Huan was able, without a show of force, to assemble the feudal lords nine times. Such was his *ren*."
>
> Zi Gong said, "I don't suppose Guan Zhong was a man of *ren*. Not only did he not die for Prince Jiu, but he lived to help Huan who had the Prince killed." The Master said, "Guan Zhong helped Duke Huan to become the leader of the feudal lords and to save the Empire from collapse. To this day, the common people still enjoy the benefit of his acts. Had it not been for Guan Zhong, we might well be wearing our hair down and folding our robes to the left. Surely his was not the petty faithfulness of the common man or woman who commits suicide in a ditch without anyone taking any notice" (*Analects* 14:16-17, Lau (trans.) modified).

In other words, both disciples felt that Guan Zhong's conduct was contrary to *ren* because he not only did not die with his master as his colleague did, he even served the murderer of his master. They considered him an obvious example of "staying alive at the expense of *ren*"; he should have committed suicide in order to preserve *ren*, which was required by Confucian Thesis I. Confucius, however, defended Guan Zhong as a man of *ren* because (1) he brought about a peaceful coexistence of all the vassal lords and prevented warfare among them, (2) by helping Duke Huan to become the leader of the vassal lords he saved the Empire from collapse, and so prevented the imminent invasion and conquest of the barbarians, and (3) in virtue of his enormous contribution to national interests he should be exempted from the petty morality of the common people.

Confucius' apparent exemption of Guan Zhong from the requirement of Confucian Thesis I is intriguing. In spite of his moral authority, many subsequent followers of Confucius still could not appreciate Guan Zhong and found his refusal to die with his master a serious character flaw. In the history of interpretation of the *Analects*, one can find that though many commentators appreciated Confucius' defense of Guan Zhong, many others were confused by their Master. They therefore developed an ingenious exegesis so as to understand Confucius as either not appreciating Guan Zhong at all, or have mixed feelings about Guan Zhong's moral merit.[24] For example, the eminent neo-Confucian philosopher Cheng I (1033-1107) changed the birth order of the brothers. He opined that Duke Huan must be the elder brother and Prince Jiu the younger one. Accordingly, it was legitimate for the former Prince Xiaobai to succeed his father as the Duke of Qi (i.e., Duke Huan). Though the failure to die with his master was still morally wrong, Guan Zhong's assistance of Duke Huan, who was supposed to be the legitimate heir to the throne, was morally excusable. Were it the other way round, i.e., Duke Huan was the younger brother, and he had violated the order of succession by usurping the throne, it would be extremely immoral for Guan Zhong to assist him; he would have no choice but to commit suicide. However, as many subsequent commentators pointed out, Cheng I's version of birth order was entirely wrong; the reality was just the opposite. Cheng I changed the birth order because this was the only way he could make any sense out of Confucius' defense of Guan Zhong. This shows how strongly many Confucians felt that Guan Zhong should have committed suicide.

One way to understand ethically this debate on Guan Zhong's refusal to commit suicide is to note that there have been divergent understandings of the proper manifestation of *ren*. Some understood it narrowly as a norm in personal ethics only, and considered the lack of total dedication to one's master a hallmark of the violation of *ren*. Furthermore, they thought that serious character flaws such as this one could not be compensated by achievement, no matter how large, in the social-political world. This was the mainstream moral thinking of Confucianism. In the

Song Dynasty, however, there was another stream of neo-Confucianism that was known as the "Utilitarian" school, whose major representatives were Chen Liang and Ye Shi (for a helpful discussion of the former in English, see Tillman, 1982). They criticized the mainstream neo-Confucianism for over-emphasizing personal morality and neglecting social-political achievements. For them, to promote the social-political good was also a manifestation of *ren*. Chen Liang, for example, appreciated Confucius' defense of Guan Zhong very much, whereas Zhu Xi, with whom he had an extensive debate through correspondence, still had mixed feelings about Guan Zhong.[25] For Chen Liang, Guan Zhong did not stay alive in spite of Confucian Thesis I; he did so in virtue of Thesis I. *Ren* in the wider sense required him not to die, but to live. To put the point another way, one can say that those who thought that both Qu Yuan and Guan Zhong should not commit suicide subscribed to Antithesis I as follows:

> **Antithesis I: One should broaden the scope of one's commitment; instead of dying for a rather limited cause, one should live and die for an object of a higher order.**

This antithesis does not rule out other-regarding suicide in principle, but limits it to a more restricted number of circumstances.

The most eloquent exposition of Antithesis I can be found in the Confucianism of the early Qing Dynasty, especially in the ethics of the so-called "Three Great Confucians of the late Ming and the early Qing," viz., Huang Zongxi (1610-1695), Gu Yanwu (1613-1682), Wang Fuzhi (or Wang Chuanshan, 1619-1692). All three of them, in their youth, witnessed the downfall of the Ming Dynasty and the invasion of the Manchu people. They all joined the armed resistance movement in the attempt to expel the Manchus and restore the Ming Dynasty, not out of the unswerving loyalty to the royal house of Ming, but out of the mission to keep China from the domination by a foreign people and its "barbarian" culture. Their resistance was in vain and the Manchus swiftly occupied the entire China. It was an age of total chaos, and Huang Zongxi described it as "an age of the falling apart of both the heaven and the earth." Many patriotic Chinese, under the influence of Confucian Thesis I, committed suicide. These three Confucians not only did not commit suicide, but also articulated a new ethics of suicide which was an important breakthrough in Confucianism.

Both Gu Yanwu and Wang Fuzhi were deeply appreciative of Guan Zhong's career. Like Chen Liang in the Song Dynasty, they defended Guan Zhong's refusal to commit suicide and his serving Duke Huan. However, unlike Chen Liang, they did not defend him on utilitarian ground. According to them, Guan Zhong's decision to stay alive was justified by a weightier moral duty, viz., the duty to keep China free from foreign and barbarian domination. Though Guan Zhong did not discharge the duty to die with Prince Jiu, it was because this duty was "trumped" by the much more important duty to defend China and its civilization. Accordingly, even if we understand *ren* narrowly as a norm in personal ethics, *ren* manifests itself not merely as dedication to one's master, but more importantly as dedication to one's country. Though Guan Zhong did not commit a suicide of *ren* and *yi*[a], he stayed alive for a greater *ren* and *yi*[a]. Or to put it differently, though *ren* requires one to die with the object of ultimate dedication, Prince Jiu was not such an object and so *ren* did not require Guan Zhong to die with him. (Gu, 1990, p.317; Wang, 1975, pp.412-415). Hence Gu and Wang found Guan Zhong's deeds inspiring for them. Though the Ming Dynasty perished and the Manchu political domination was a *fait accompli*, they opted against dying together with Ming. They had to stay alive to defend Chinese civilization so that the Manchu conquest of China would remain only in the military and political aspects.[26]

Though Huang Zongxi did not write on this topic as much as Gu and Wang did, his view can be traced through an essay by Chen Que (1604-1677), whose view he fully shared. In the essay "On Dying for Integrity" (*Sijie Lun*) Chen argued that many people in the past have misapplied Confucian Thesis I, that "many people, out of the desire for fame, committed suicide through drowning, hanging, taking lethal drug, and cutting one's throat. They aspired to die with no regret. Innumerable sons followed their fathers to death, wives followed their husbands, and educated persons (*shi*) followed their friends. They just wanted to commit suicide without proper regard of its propriety. There were even unwed girls who died for men they secretly admired, and educated persons who died for someone unacquainted. The uneducated people reinforced this ethos through giving favorable publicity to each other's suicide. There is nothing more harmful to the customs of society than this... Since 1644 [the year the Ming Dynasty fell and the Manchus occupied the capital, Beijing] many people committed suicide as well. People say these were suicides for the sake of *yi*[a]. But it is not the right *yi*[a] to die for and a person of stature should not do it. Furthermore, whether a person is virtuous or not can be seen through his or her entire course of life... Now the focus of moral evaluation has been shifted to the manner of death, hence adulterers, robbers, actresses, and prostitutes can all be counted virtuous persons. No value has been so confusing as that in dying for integrity. How distressing it is!" (Chen, 1979, pp.153-154, trans. mine).

Huang Zongxi fully shared Chen Que's deploring of the epidemic of suicide in the name of *ren* and *yi*[a], especially those that took place immediately after the fall of the Ming Dynasty. In the long epitaph he composed for Chen, Huang quoted approvingly a section of Chen's essay "On Dying for Integrity" and made the comment that Chen's view was beneficial to social morals (Huang, in Chen, 1979, p.8). It is noteworthy that Chen Que and Huang Zongxi were fellow-students of Liu Zongzhou (1578-1645), an eminent late Ming Confucian philosopher who died for integrity after the fall of the Ming Dynasty. Chen and Huang not only did not follow the example of their teacher, whom they admired greatly in other respects, but also deplored the excess of dying for integrity in society. This was an important breakthrough.[27]

It should be noted that these three Confucians did not reject dying for integrity altogether. They only intended to modify Confucian Thesis I by broadening the object of dedication from an emperor, a royal house, and a dynasty to the entire country. They resolved to stay alive as long as they could still contribute to China in some way (e.g., to preserve her civilization) and to commit suicide only if they were compelled to do otherwise. (Hence Gu threatened to commit suicide when the Qing central government repeatedly asked him to enter public life and serve the Manchu regime.) Their ethics of suicide is the best exposition of Confucian Antithesis I.

Another way to analyze ethically the divergent evaluations of the suicide of Qu Yuan and the refusal to commit suicide of Guan Zhong is to make use of the idea of *san buxiu*, i.e., "Three Forms of Immortality," viz., to establish virtue in personal morality, to establish successful public service, and to establish speech/writing (*Zuo Zhuan*, or *Zuo Qiuming's Commentary on the Spring and Autumn Annals*, Twenty-fourth year of Duke Xiang). If one can establish one of the three lasting influences, even if one died, one would be immortal. The knowledge of these "three forms of immortality" was widespread in ancient China. To commit an other-regarding suicide, as Qu Yuan did and Guan Zhong should have, is to establish virtue in personal morality. To refuse to commit suicide, as

Guan Zhong did and Qu Yuan should have, is to establish successful public service. For those people who think the first form of immortality is of the utmost importance, they will commend other-regarding suicide, and so praise Qu Yuan and condemn Guan Zhong. For those who think the second form of immortality is at least of equal importance, however, they will commend a refusal to commit suicide so that one can provide successful public service to the country; they will commend Guan Zhong and lament over Qu Yuan. (Accordingly, Sima Qian's decision of refusing suicide can be understood as an attempt to establish speech/writing over personal virtue.) In other words, the first form of immortality is consonant with Confucian Thesis I and II, which are more favorable to suicide. The second and third forms of immortality, however, are more consonant with Antithesis II, which is less favorable to suicide. How are we to weigh these three forms of immortality in case they come in conflict? This is an issue that contemporary Chinese ethics should address.

VII. IMPLICATIONS FOR EUTHANASIA

How do the ancient Confucian ethical views on suicide bear on the contemporary bioethical issue of euthanasia?[28] I have seven brief observations.

First, "euthanasia" in the etymological sense of "good death" is agreeable with Confucian Thesis I. Sima Qian's point in distinguishing deaths that are weightier than Mount Tai from deaths lighter than swan's down is precisely to distinguish good deaths from bad or valueless deaths. Wen Tianxiang's poem also points out that one should choose a preferable way of dying. Confucianism never had a doctrine of the sanctity of human life, and never deemed suicide intrinsically wrong. In some circumstances, death by suicide is a better death than a "natural" death. However, we should also note that a "good death" in Confucianism does not mean a dying process which is swift, peaceful, and free of pain, but a death for the sake of *ren* and *yi*[a], a death that can render service to others, or a death that expresses abiding dedication to others. In short, a "good death" in Confucianism is good for other-regarding reasons, rather than for self-regarding reasons as contemporary proponents of euthanasia understand it.

Second, Confucian Thesis I justifies altruistic suicides. Can we therefore infer that it also justifies altruistic euthanasia, i.e., to request euthanasia for the sake of relieving the burden (emotional, financial, and otherwise) to others (family and society)? It is not clear to me that Confucian Thesis I would apply in these circumstances. This is because altruistic suicides in ancient China were usually intended to render a positive benefit to others. Altruistic suicides in the form of negatively removing the "burden" to one's family and to society were not encouraged.[29] This is because according to the Confucian social vision, the good society is one in which, through an extensive support network, "widows, widowers, orphans, the childless, the disabled, and the sick are to be well taken care of."[30] Furthermore, the elderly were highly esteemed in ancient China; they were considered "senior citizens" in the literal and positive sense. Hence the sick elderly were not allowed to be seen as "burden" to anyone. If altruistic suicide for the sake of relieving the burdens of others was to be discouraged, altruistic euthanasia for the same reason would be discouraged as well.

Third, Confucian Thesis II seems to endorse euthanasia. Both have to do with terminating life for a self-regarding reason, and both can be characterized as "death with dignity." Indeed, there are certain parallels between the ancient Chinese understanding of death with dignity and the contemporary bioethical understanding of death with dignity. Regarding type I of death with dignity in ancient China (see Section III above), the parallels between the Chinese and the contemporary understandings are as follows:

(1) Death is imminent.

(2) The manner of death is highly likely to be undignified because of external factors. For example, (a) in ancient China it is a humiliating death to be executed by one's enemies, by the emperor, or by the government; (b) in contemporary bioethics it is impersonal or excessive medical intervention that can lead to an undignified death, e.g., machines, tubes, and an over-zealous medical staff. ("[A] medicalization and institutionalization of the end of life that robs the old and the incurable of most of their autonomy and dignity: Intubated and electrified, with bizarre mechanical companions, confined and immobile, helpless and regimented, once proud and independent people find themselves cast in the role of passive, obedient, highly disciplined children" (Kass, 1991, p.132).)

(3) Suicide is committed as a way of escape from undignified treatment and a way to preserve one's dignity.

Such parallels notwithstanding, it seems to me that there are also certain dissimilarities between these two understandings.

(1) The assault on human dignity in the Chinese understanding is entirely external (from enemies, emperor, government) and also not universal (restricted largely to warriors, rebellious generals, and government officials). The assault on human dignity in the issue of euthanasia is partly internal (disease, old age, bodily and mental decay all stem from our mortal and corruptible body) and is a universal human phenomenon.
(2) In the Chinese understanding, captivity is unavoidable and so is the subsequent humiliation; one's destiny is controlled by hostile forces, and nobody can help one to ameliorate one's expected suffering. In the issue of active, voluntary euthanasia, one's suffering, at least in some cases as it was argued, can be reduced by palliative care; we are not captured and isolated in a maleficent environment, but are surrounded by health care professionals who are, supposedly, there to help us.

Regarding type II of death with dignity in ancient China, there are also parallels and dissimilarities between the two understandings. The parallels are:

(1) Death is not imminent. Though one's life is not threatened, one decides that one is better off dead than alive.
(2) Suicide is committed as a way of avoiding humiliation and thus preserving one's dignity. In the Chinese understanding the indignity stems from the trial in court and/or the imprisonment; in the issue of euthanasia the indignity stems from an incurable illness.

And there is also one dissimilarity: In the Chinese understanding the source of indignity is external and circumstantial (viz., a legal system), whereas in the issue of euthanasia the source of indignity is intrinsic to our mortal embodied life. Hence for the latter, the indignity is part of the human condition; the same thing, however, cannot be said of the former. In short, the contemporary understanding of death with dignity, viz., active, voluntary euthanasia, in most cases is not a close analogue of death with dignity in ancient China.

My fourth observation is related to the third. Just as Confucian Thesis II was countered by Antithesis II, likewise, euthanasia, at least in the non-terminal context, will be opposed by Antithesis II. This is because, according to this antithesis, in spite of personal tragedy and living in undignified circumstances, one should embark on or continue one's project in life so as to make the most out this life. Sima Qian went through excruciating anguish and tormenting mental distress, but he resisted the invitation to die.[31]

Fifth, in a terminal context, what about suicide and euthanasia for the sake of relieving one's own intractable and end-of-life suffering? When there is no more time and energy for one to engage in a project, what is the purpose of enduring the intractable pain? Is euthanasia in this context morally permissible according to Confucian values? I do not think there is an obvious answer to this question. For one thing, historically Confucianism has been more concerned with teaching people to cultivate one's quality of moral life than with teaching people to maintain the quality of biological life. For another, as Mencius argues, compassion is the germ of *ren* or benevolence (*Mencius*, 2A:6), and so we cannot be indifferent to others' intense suffering. Compassion dictates that we should devote more resources to palliative medicine and make hospice care more easily available. If, however, all palliative treatments fail, and terminating one's life immediately is the only way of relieving one's dragged out suffering, I do not see how euthanasia is opposed by *ren*. In short, compassion and benevolence can be expressed in different ways, depending on the variation of empirical factors such as the availability and effectiveness of palliative treatments.

Sixth, for many contemporary proponents, euthanasia has more to do with individual autonomy than with the relief of pain. As an article in the *New York Times* (August 14, 1991, A19) puts it, "Pain management and hospice care are better than ever before. But for some people they are simply the trees. The forest is that they no longer want to live, and they believe the decision to die belongs to them alone" (quoted from New York State Task Force on Life and the Law, 1994, p.87). Confucianism is not unsympathetic to the idea that a moral agent should have some control over the time and circumstance of his or her death. Confucian Theses I and II certainly grant, and even encourage, individual autonomy in deciding the time and circumstance of one's death. This favorable inclination notwithstanding, Confucianism does not regard such autonomy as open-ended. The Confucian freedom to die is not without boundary (i.e., to do as one wishes), but is to be guided by *ren* and *yi*[a] (i.e., to do as one ought to). The Confucian freedom to die is not valuable in itself, but only as a means to serve a moral cause. "I do not want to live in this condition, period" is not an acceptable Confucian reason for euthanasia. Consequently, autonomy in dying is not a private or strictly personal issue, none of others' business. On the contrary, if autonomy in dying is to be guided by *ren* and *yi*[a], it is a moral and public issue – a matter of public concern, though not necessarily of public interference.

Seventh, this paper is limited to an analysis of the Confucian moral evaluation of suicide, and does not touch on public policy issues. Hence the *moral* conclusions above do not have a direct bearing on *legal* issues such as the legalization of euthanasia and of physician assisted suicide, which involve issues not examined in this article, e.g., the role and responsibilities of physicians, probable societal consequences (intended and unintended) of such changes in law, potential for abuse, etc.[32]

Department of Religion and Philosophy, and
Center for Applied Ethics
Hong Kong Baptist University
Kowloon Tong, Hong Kong

NOTES

1 Most of the debates on his suicide, with a span of more than sixty years, are now conveniently collected in Luo, 1995. For a brief account in English of the circumstances of Wang's suicide, see Bonner, 1986, pp.206-215.

2 For an ethical discussion on the difference between human acts and bare events, see Donagan, 1977, pp.37-52.

3 For more detailed discussions on various types of suicide in ancient China, see Eberhard, 1967; Guo, 1989; He, 1993; Hsieh and Spence, 1981; Huang and Wu, 1992; Lau, 1989; Lin, 1990; Lin, 1992; Maclagan, 1908; T'ien, 1988; Zhang, 1993.

4 Altruistic or other-regarding suicide was relatively rare in the West; hence Durkheim says that it is frequent only in "lower societies" (Durkheim, 1951, p.217).

5 Formally speaking, the major moral issue in the premodern West and in premodern China was the same, viz., "Is it morally permissible not to perform a particular duty?" The issue was different only substantively. In the premodern West, the duty in question was the duty not to commit suicide, whereas in premodern China, the duty in question was the duty to commit suicide.

6 Mencius is to Confucius what Paul is to Jesus in Christianity.

7 I found these sayings in G. Chen, 1990, pp.559-564, under the heading of "To lay down one's life for a cause of *yi*[a]."

[8] Mount Tai was a sacred and famous mountain in ancient China in the Shandong province.

[9] The sources of my information are from *Shiji* (*Records of the Historian*) and some collections of Chinese novels in the Ming Dynasty. The former contains the biographies of many celebrities in ancient China whereas the latter narrates the stories and legends of many ordinary people in medieval China.

[10] It seems to me that the western philosopher that comes closest to this classical Confucian view is Immanuel Kant. In his lecture on suicide he emphasizes repeatedly that "life is not to be highly regarded for its own sake. I should endeavour to preserve my own life only so far as I am worthy to live ... Yet there is much in the world far more important than life. To observe morality is far more important. It is better to sacrifice one's life than one's morality. To live is not a necessity; but to live honourably while life lasts is a necessity" (Kant, 1930, pp.150-152). Accordingly, though Kant firmly opposes suicide in the sense of self-destruction, he commends self-sacrifice highly. Risking one's life and willing to be killed for the sake of others' good are praiseworthy. Furthermore, altruistic suicide, i.e., actively to kill oneself for others' sake, is also noble, as in the example of Cato the Younger (95-46 BC), who "knew that the entire Roman nation relied upon him in their resistance to Caesar, but he found that he could not prevent himself from falling into Caesar's hands. What was he to do? If he, the champion of freedom, submitted, every one would say, 'If Cato himself submits, what else can we do?' If, on the other hand, he killed himself, his death might spur on the Romans to fight to the bitter end in defense of their freedom. So he killed himself. He thought that it was necessary for him to die. He thought that if he could not go on living as Cato, he could not go on living at all. It must certainly be admitted that in a case such as this, where suicide is a virtue, appearances are in its favour" (Kant, 1930, p.149). This high regard for altruistic suicide notwithstanding, one should not overlook that immediately after the aforequoted passage, Kant cautions, "But this is the only example which has given the world the opportunity of defending suicide. It is the only example of its kind and there has been no similar case since"(*ibid.*).

[11] As in *zijin, zicai, zijing, ziqiang, qingsheng*

[12] As in *xunguo, xunjun, xunzhu, xunfu, xunjie, xunqing, xunzang, xunsi, xundao.*

[13] As in *sijie, qijie, shijie, xunjie, jielie, jiecao.*

[14] For an assessment of the suicide of the former, see Lin, 1976; for debates on the suicide of the latter, see Luo, 1995.

[15] Chen's suicide is an admonishing suicide (*sijian*), which has a long history in China, dated back to Qu Yuan (see section V below). It can be compared to suicide as social protest in the West (cf. Battin, 1996, pp.92-93).

[16] Cf. Battin, 1996, pp.67-68.

[17] As a famous line from chapter one of *Xiaojing* (*Book of Xiao*) goes, "Our body, limbs, hair, and skin all originated from our parents. We should hold them in respect and guard them against injury. This is the beginning of filial piety."

[18] It should be noted that though Daoism and Buddhism were not established religions, they flourished in Chinese society. The persecution of non-established religions and ideologies occurred only infrequently in China.

[19] These two passages are taken from *Liji* (*The Book of Rites*).

[20] For stimulating discussions on the suicide of these two intellectuals, see Huang, 1989; Wang, 1986; Chen, 1986; Su, 1986.

[21] Again, Kant's idea on self-regarding duty comes very close to Confucian Thesis II. "We are in duty bound to take care of our life; but in this connexion it must be remarked that life, in

and for itself, is not the greatest of the gifts entrusted to our keeping and of which we must take care. There are duties which are far greater than life and which can often be fulfilled only by sacrificing life ... If a man cannot preserve his life except by dishonouring his humanity, he ought rather to sacrifice it; ... It is not his life that he loses, but only the prolongation of his years, for nature has already decreed that he must die at some time; what matters is that, so long as he lives, man should live honourably and should not disgrace the dignity of humanity ... If, then, I cannot preserve my life except by disgraceful conduct, virtue relieves me of this duty because a higher duty here comes into play and commands me to sacrifice my life" (Kant, 1930, pp.154-157). Accordingly, Kant thinks that in the case of an innocent man wrongly accused of treachery, if he is given the choice of death or penal servitude for life, he should choose the former. Similarly, a woman should prefer to be killed to being violated by a man. Kant, however, stops short of recommending suicide in order to avoid such dishonor. Battin therefore points out, correctly, that Kant is inconsistent here. If our self-regarding duty of avoiding moral degradation is of such paramount importance, "and if death – the only possibility for nondegradation – is the only morally acceptable alternative, the only way to achieve this alternative would be to take death upon oneself" (Battin, 1996, p.109). In other words, as an eminent contemporary Kantian scholar argues, the spirit of Kant's ethics should permit some self-regarding suicides (Hill, 1983).

[22] For the Chinese texts, see D. Shi, H. Shi, Z. Shi; for a helpful general discussion, see Jan, 1964.

[23] For a general account of Qu Yuan's life and significance in English, see Schneider, 1980.

[24] For an extensive samplings of such imaginative exegesis, see Cheng, 1990, pp.981-996.

[25] The correspondence between Chen Liang and Zhu Xi were all collected in *Chen Liang Ji*, or *Writings of Chen Liang*, 1987, pp.332-76.

[26] It should be noted that this broader notion of loyalty and suicide for its sake was advanced outside Confucianism a long time ago. Since Guan Zhong was such a legendary hero, and was widely referred to in the literature of late Warring States period, shortly after his death there was an oral tradition of thought that was dedicated to him. This stream of thought was subsequently crystallized in a collection of writings that bore his name, viz., *Guan Zi*. In Chapter XVIII of this work ("Dakuang") Guan Zhong was supposed to have said, "As an official to the monarch, I shall carry out the mandate of the monarch, to serve the state and the ancestral shrine of the ruling house. How can I die just for Prince Jiu alone? If the state perishes, the ancestral shrine of the ruling house is destroyed, sacrificial offering stops, I shall then follow to death. I shall stay alive for any disaster short of these three. As long as I am alive, Qi will be benefited; if I die, it will be a loss to Qi" (trans. mine).

[27] Wang Fuzhi, in his commentary on chapter eight of the writings of Zhuang Zi, also expressed his agreement to a large extent of the Daoist critique of suicide for the sake of *ren* and *yi*[a] (see Wang, 1977, pp.76-81).

[28] In this section the term "euthanasia" is confined to voluntary, active euthanasia only.

[29] For a helpful discussion on the distinction between negative and positive altruistic suicide, see Battin, 1996, pp.84-93.

[30] This is a famous passage from chapter 9 ("Liyun") of *Liji* (The *Book of Rites*), an important Han Confucian canon.

[31] As a very famous passage from *Mencius* (6B:15, Lau (trans.) modified) goes, "That is why Heaven, when it is about to place a great task on a man, always first tests his resolution, exhausts his frame and makes him suffer starvation and hardship, frustrates his efforts so as to shake him form his mental lassitude, toughen his nature and make good his deficiencies."

[32] Subsequent to the writing of this article, I have elaborated some parts of this article into two lengthy papers; one is already published and the other is forthcoming: (1) "Confucian Values of Life and Death & Euthanasia" (in Chinese), *Chinese and International Philosophy of Medicine* 1:1 (Feb 1998):35-73; (2) "Confucian Ethic of Death with Dignity and Its Contemporary Relevance," *The Annual of the Society of Christian Ethics*, volume 19 (1999): forthcoming. The research for these two papers and the present article was funded by Hong Kong Baptist University Faculty Research Grant (FRG/94-95-II-04) and by Research Grants Council of Hong Kong 1995-1996 Earmarked Research Grant (RGC/95-96/26), and was assisted by Miss Lee Wing Yi. The present author is grateful to them for their respective contribution.

REFERENCES

Battin, M. P.: 1996, *The Death Debate: Ethical Issues in Suicide,* Prentice-Hall, Upper Saddle River, New Jersey.

Bonner, J.: 1986, *Wang Kuo-wei: An Intellectual Biography*, Harvard University Press, Cambridge, Massachusetts.

Chen, C.: 1986, 'Si: gei "wenge" (Death: For the 'Cultural revolution'),' *Shanghai wenxue (Literature in Shanghai)* 9: 4-11.

Chen, G., *et al.*: 1990, *Zhongguo gudai mingju cidian (Dictionary of Famous Sayings in Ancient China))*, Shanghai cishu Press & Wanli shudian, Shanghai.

Chen, L.: 1987, *Chen Liang Ji (Writings of Chen Liang),* Zhonghua Press, Beijing.

Chen, Q.: 1979, 'Sijie Lun (On Dying for Integrity)', in *Chen Que Ji (Works of Chen Que)*. Two volumes. Zhonghua Press, Beijing, pp.152-155.

Chen, T.: 1982, 'Jueming Ci (Suicide Note)', in Liu Qingpo and Peng Guoxin (eds.), *Chen Tianhua Ji (Works of Chen Tianhua)*, Hunan renmin Press, Changsha.

Cheng, S.: 1990, *Lunyu Jishi (Collected Commentaries on the Analects),* Four volumes, Zhonghua Press, Beijing.

Chi, C.: 1972, 'The Orphan of Chao,' in Liu Jung-en (trans.), *Six Yuan Plays,* Penguin Books, New York, pp.41-81.

Chuang, T.: 1964, *Chuang Tzu: Basic Writings,* Burton Watson (trans.), Columbia University Press, New York.

Confucius: 1992, *Analects,* D. C. Lau (trans.), second edition, Chinese University Press, Hong Kong.

Dayal, H.: 1932, *The Bodhisattva Doctrine in Buddhist Sanskrit Literature,* Routledge & Kegan Paul, London.

De La Vallee Poussin, L.: 1908, 'Suicide (Buddhist)' in *Encyclopedia of Religion and Ethics,* James Hastings (eds.), volume 12, T & T Ckark, Edinburgh, pp.24-26.

Donagan, A.: 1977, *The Theory of Morality,* University of Chicago Press, Chicago.

Dong, Z.: *Chunqiu Fanlu (Exuberant Dew of the Spring and Autumn),* many editions.

Durkheim, E.: 1951, *Suicide: A Study in Sociology: A Study in Sociology*, John A. Spaulding and George Simpson (trans.), Free Press, New York.

Eberhard, W.: 1967, 'Suicide in short stories,' in *Guilt and Sin in Traditional China*, University of California Press, Berkeley.

Gu, Y.: 1990, *Rizilu (Daily Additions to Knowledge)*, in Huang, R.:1990.

Guan Zi (*Master Guan*), many editions.

Guo, D.: 1989, *Dongfang siwang lun* (*Eastern Views on Death*), Liaoning Jiaoyu Press, Shenyang.

He, X.: 1993, *Zhongguoren de siwang xintai* (*The Chinese Mentality of Death*), Shanghai Wenhua Press, Shanghai.

Hill, T. E.: 1983, 'Self-regarding suicide: A modified Kantian view,' in *Suicide and Ethics: A Special Issue of Suicide and Life-Threatening Behavior,* Margaret P. Battin and Ronald W. Maris (eds.), Human Sciences Press, New York, pp.254-275.

Hsia, A.: 1988, '*The Orphan of the House Zhao* in French, English, German, and Hong Kong Literature,' *Comparative Literature Studies,* 25:4, 335-351.

Hsieh, A. C. K. and Spence, J. D.: 1981, 'Suicide and the family in pre-modern Chinese society,' in *Normal and Abnormal Behavior in Chinese Culture,* Arthur Kleinman (eds.), Holland/ Boston:U. S. A., Dordrecht, pp.29-46.

Huang, J. and Wu, G.: 1992, 'Gudai zhongguoren de jiazhiguan: jiazhi quxiang de chongtu ji qi jiexiao (Ancient Chinese Values: Conflict and Resolution of Value Orientation),' in *Zhongguoren de jiazhiguan guoji yantaohui lunwenji* (*Proceedings of International Conference on Chinese Values*), Hanxue yanjiu zhongxin (eds.), Hanxue yanjiu zhongxin, Taipei.

Huang, R.: 1990, *Rizhilu Jishi* (*Collected Commentaries on Daily Additions to Knowledge*), Two volumes, Huashan wenyi Press, Shijiazhuang.

Huang, Z.: 1987, *Quyuan wenti lunzheng shigao* (*A History of Debate on Qu Yuan*), Shiyue wenyi Press, Beijing.

Huang, Z.: 1989, 'Qiangu jiannan wei yisi: tan jibu xie Lao She, Fu Lei zhi si de xiaoshuo (Death as a Difficult Task since Time Immemorial: On Some Novels on the Death of Lao She and Fu Lei),' *Dushu* (*Reading*) 4 (April), 53-63.

Huang, Z.: 1979, 'Chen Qianchu xiansheng muzhiming (Epitaph for Master Chen Qianchu [i.e., Chen Que]),' in Chen, Q.:1979, pp.4-9.

Hume, D.: 1965, 'Of suicide,' in *Hume's Ethical Writings,* Alasdair MacIntyre (eds.), Collier Books, London, pp.297-306.

Jan, Y.: 1964, 'Buddhist self-immolation in medieval China,' in *History of Religions.* 4:2, 243-268.

Jia, Y.: *Diao Qu Yuan fu* (*Elegy to Qu Yuan*), in *Shiji*, by Sima Qian, Biography number 84.

Kant, I.: 1930, *Lectures on Ethics,* Louis Infield (trans.), Methuen & Co., London.

Kass, L.: 1991, 'Death with dignity and the sanctity of life,' in *A Time to be Born and a Time to Die*, Barry S. Kogan (eds.), Aldine de Gruyter, New York, pp.117-145.

Lau, J. S. M.: 1988-1989, 'The courage to be: Suicide as self-fulfillment in Chinese history and literature,' *Tamkang Review,* 19:1-4 (Autumn 1988- Summer 1989), 715-734.

Liang, Q.: 1927, 'Wang Jingan xiansheng muqian daoci (A Memorial Speech at the Grave of Mr. Wang Jingan [Wang Guowei]),' reprinted in Luo, J., 1995.

Liji, (*Book of Rites*), many editions.

Lin, Y.: 1976, 'The suicide of Liang Chi: An ambiguous case of moral conservation,' in *The Limit of Change: Essays on conservative Alternatives in Republican China,* Charlotte Furth (eds), Harvard University Press, Cambridge, pp.151-168.

Lin, Y.: 1990, *The Weight of Mt. T'ai: Patterns of Suicide in Traditional Chinese History and Culture,* Ph. D. Dissertation, University Microfilms International, Ann Arbor, Michigen.

Lin, Y.: 1992, 'Maishen maide qiannianming – lun zhongguoren de zisha yu minyu (To Acquire Immortal Fame at the cost of One's Life – On Chinese Suicide and the Longing for Famem,'

Zhongguo wenzhe yanju jikan (*Research Journal on Chinese Literature and Philosophy*), 2 (Feb): 423-451.

Luo, J.(ed.): 1995, *Wang Guowei zhi si* (*On the Death of Wang Guowei*), Qiling Press, Taipei.

Maclagan, P. J.: 1908, 'Suicide (Chinese)' in *Encyclopaedia of Religion and Ethics,* James Hastings (eds.), volume12, T & T Clark, Edinburgh.

Mencius: 1984, *Mencius*, two volumes, D.C. Lau (trans.), Chinese University Press, Hong Kong.

New York State Task Force on Life and the Law: 1994, *When Death is Sought: Assisted Suicide and Euthanasia in the Medical Context*, New York State Task Force on Life and the Law, New York.

Nivison, D. S.: 1987, 'Jen and I,' in *Encyclopedia of Religion*, Micea Eliade (eds.), volume 7, Macmillan, New York.

Schneider, L. A.: 1980, *A Madman of Ch'u: The Chinese Myth of Loyalty and Dissent,* University of California Press, Berkeley.

Schopenhauer, A.: 1962, 'On suicide,' *The Essential Schopenhauer* (English selections of *Paralipomena*), Allen & Unwin, London, pp.97-101.

Seneca: 1920, *The Epistles of Seneca,* William Heinemann, London, pp.56-73.

Shi, D.: *Xu gaoseng zhuan* (*More Biographies of Eminent Monks*), *yiishen pian* number 7, in *Dazheng xinxiu dazangjing* (*Taisho Edition of the Tripitaka in Chinese*), number 2060, pp.678-684.

Shi, H.: 1986, 'Lun fojiao de zishaguan (Buddist Views on Suicide),' *Taida zhexue lunping* (*Philosophical Review of National Taiwan University*) 9(Jan), 181-196.

Shi, H.: *Gaoseng zhuan* (*Biographies of Eminent Monks*), *wangshen pian* number 6. in *Dazheng xinxiu dazangjing* (*Taisho Edition of the Tripitaka in Chinese*), number 2059, pp.403-406.

Shi, Z.: *Song gaoseng zhuan* (*Biographies of Eminent Monks in Song Dynasty*), *yishen pian* number 7, in *Dazheng xinxiu dazangjing* (*Taisho Edition of the Tripitaka in Chinese*), number 2061, pp.855-862.

Sima, Q.: 1965, 'Bao Ren An Shu (Letter to Ren An),' For a free translation, see *Anthology of Chinese Literature: From Early Times to the Fourteenth Century*, Cyril Birch(eds.), Grove Press, New York, pp.95-102.

Sima, Q.: 1961, *Shiji, (Records of the Historian),* many editions, for an abridged translation, see *Records of the Grand Historian of China*, two volumes, Burton Watson (trans), Columbia University Press, New York.

Song S. (*History of Song Dynasty*), Biography number 177, many editions.

Su, S.: 1986, 'Lao She zhi si (Death of Lao She),' *Renmin wenxue* (*People's Literature*) 8:22-34.

T'ien, J.: 1988, *Male Anxiety and Female Chastity: A Comparative Study of Chinese Ethical Values in Ming-ch'ing Times,* E. J. Brill, Leiden.

Tang, J.: 1974, 'Zhongguo wenhua yu shijie (A Manifesto for a Re-appraisal of Sinology and Reconstruction of Chinese Culture),' *Shuo zhonghua minzu zhi huaguo piaoling* (*On the Diaspora of the Chinese*), Sanmin Press, Taipei.

Tillman, H. C.: 1982, *Utilitarian Confucianism: Cheng Liang's Challenge to Chu Hsi*, Harvard University Press, Cambridge, Massachusetts.

Wang, F.: 1975, *Du sishu dachuan shuo* (*Discussions After Reading the Great Collection of Commentaries on the Four Books*), Two volumes, Zhonghua, Beijing.

———.: 1977, *Zhuang Zi Jie* (*An Interpretation of Zhuang Zi*), Zhonghua Press, Hong Kong.

Wang, Z.: 1986, 'Bayue jiaoyang (Bright Sunshine in August),' *Renmin wenxue* (*People's Literature*) 9:17-21.

Xiaojing (*Book of Filial Piety*), many editions.

Yang, X.: *Fan Li Sao, (Antithesis to The Lament),* in Zhu Xi, *Chuci jizhu.*
Zhang, S.: 1993, *Siwang zhi si yu siwang zhishi* (*Thoughts and Poems on Death*), Huazhong Ligong University Press, Wuchang.
Zheng, X.: 1994, *Zhongguo siwang zhihui* (*Chinese Wisdom on Death*), Dongda Press, Taipei.
Zhu, X.: *Chuci jizhu* (*Collected Commentaries on Chu Poetry*), many editions.
Zuo, Q.: *Zuo Chuan* (*Commentary on the Spring and Autumn Annals*), many editions.

GEORGE KHUSHF

REFLECTIONS ON THE DIGNITY OF GUAN ZHONG: A COMPARISON OF CONFUCIAN AND WESTERN LIBERAL NOTIONS OF SUICIDE

I. INTRODUCTION

In "Confucian Views on Suicide," Ping-Cheung Lo (1999) provides much more than a simple survey of what the Confucian tradition has to say about suicide. Like Emile Durkheim's *Suicide: A Study in Sociology (1951)*, Lo's discussion probes the deeper issues of human identity and social structure that come to expression in debate about suicide. His essay thus reaches far beyond the topic. It can be read as a more general overview of central themes in the Confucian tradition. Further, the essay conveys the force of a living tradition, and does it in a way that challenges Western sensibilities. Confucian thought emerges as a viable alternative to liberal thought. In this response, I consider the contrast between Eastern and Western political thought, and how the topic of suicide relates to others such as individual rights, social responsibility, and the scope of government. My goal is to extend Lo's analysis to a broader comparison between Western liberalism and Confucian moral and political theory.

The deficiency of a certain kind of liberalism is apparent from the case studies that become the focus of much of Lo's reflection. In the West, the morality of suicide has become an important issue in cases where a patient seeks release from the suffering that attends an incurable disease. The question is whether such patients are still bound to the rule not to take one's own life (the legal prohibition against suicide). Lo defines this as "self-regarding suicide," because the people who want to commit suicide do it out of regard for their self, viz., their desire to escape the suffering that attends terminal disease and control the time and manner of their own death.

The Second Court of Appeals in the United States recently considered the claims of three people who wanted a protected right to assisted suicide (Quill v. Vacco, 1996). Their accounts are representative. "Jane Doe" suffered from a cancerous tumor in her neck; George Kingsley and William Barth had AIDS. In their testimony, they state how they are

Ruiping Fan (ed.), Confucian Bioethics, 103–125.

suffering, and, at the stage when "suffering becomes intolerable," they would like a physician to assist them in death. Advocates of suicide claim that such an act should be regarded as private, and thus a government which acknowledges liberty and pluralism should in no way interfere. They argue that it would be oppressive to use law to force people to endure such needless suffering. The constraints placed by the state on what a person can do to self should thus be relaxed at the end of life, when a person has a terminal disease. Proponents of legalized suicide thus align decisions about suicide with the protected realm of individual liberty (and privacy), and they appeal to compassion and pity as a basis for allowing suffering individuals to end their life. Western arguments for legalization thus focus on self-regarding suicide and presuppose liberal notions of liberty, limited government, and the distinction between law and morality.

In the Confucian tradition, a very different type of case comes to the fore. Lo opens his essay with the case of Wang Guowei, a professor at the Quing Hua University, who drowned himself. Wang did not seek in this act to escape suffering; to the contrary, he sought to inspire others to moral action, so that they would take up the cause of China. Wang was concerned about the moral laxity and self-indulgence that came with the Western influence on China. His suicide can thus be read as a reaction against the self-regarding focus that is emphasized in Western attempts to legalize suicide.

The precedent for this "other-regarding suicide" is found in the primary literature of the Confucian tradition, as seen, for example, in Qu Yuan, whose suicide is celebrated in the Duan Wu Festival. Unlike the paradigmatic Western cases, the Confucian ones involve individuals who, when confronted with impossible situations, do not shrink from even the most demanding moral obligation. The difficulty, hardship, and impossibility of their situation does not justify a reduced account of obligation, as in the Western cases. They call for a heightened demand, whose fulfillment involves a final act of self-transcendence, whereby an accountability to others and, even more importantly, to the moral principles of *ren* and *yi*[a] (humanity and justice) is central. The dignity of such cases provides a marked contrast with the indignity of so-called "death with dignity" in the West.

Lo's contrast, seen in the difference between the self-regarding and other-regarding cases that become the focal point of the discussion, challenges the way the Eastern view of suicide has traditionally been understood in the West. Normally, the Chinese view of suicide is associated with a shame culture, and the act of suicide is viewed as self-regarding; namely, as a way of saving face. It is interesting to note that Lo says little about the word "shame" in his essay. While a more detailed account probably should involve more on the notion of shame and show how the other-regard is manifest as a self-judgment that depends on communal assessment of individual behavior, Lo provides a valuable service by highlighting the deeper concerns with humanity and justice (*ren* and *yi*[a]) that motivate the Confucian understanding of suicide.

From a Western perspective, there are both positive and negative characteristics of the Confucian position, and these both should be directly addressed in any constructive comparison. Positively, the moral courage and responsibility of the individual who commits suicide should be appreciated, especially in comparison with the Western cases, where the candidates for suicide evoke pity, rather than emulation. There is something genuinely inspiring in the Eastern example. On the other hand, a system where it may be obligatory for an individual to take his or her own life, especially for the cause of shame or another's welfare, seems dark and dangerously close to losing a proper sense of the majesty and dignity of each individual. It seems to allow for an undue influence of society (with its rituals) and social pressure over the individual. When set in the feudal Confucian framework, the obligation to commit suicide often seems petty and shallow. If one notes the exceptions to such obligation that can be made for those near the apex of the hierarchy (e.g., Guan Zhong), then the system seems to involve a double standard, which is elitist and ignores principles of human equality that are central for the West. These will be important considerations, when evaluating the Confucian position as a viable option, rather than simply as a tradition of historical interest.

In the end, I argue in this essay for a variant of the liberal tradition, claiming that it captures something central about human dignity, something that is missing in the Confucian tradition. However, I will not argue for the kind of liberalism that seeks to legalize suicide. Western debate about suicide is often viewed as a simple contrast between liberals, who want to legalize it, and illiberal conservatives, who seek to legislate moral principles and thus turn to force and law, rather than virtue and individual liberty. I argue for a third alternative to that debate. I distinguish between two significantly different strands of liberal thought, and argue for a version that links the prohibition of suicide with the notion of an inalienable right. It will be found that the limits on government, and thus the moral/legal distinction so important to liberalism, is generated by a notion of human dignity that entails limits on what individuals can do with themselves. From this idea, limits are generated on what society and the state can do to individuals. In the context of this approach to liberalism, a different kind of case example comes to the fore, one that is more favorably compared to the noble examples found in the Confucian tradition. Further, I suggest that one finds such an example of human dignity in the Confucian tradition itself; namely, in the case of Guan Zhong.

II. ON THE DISTINCTION BETWEEN LAW AND MORALITY

At the end of his essay, after reviewing the Confucian position on suicide in some detail, Lo qualifies his account in a way that initially struck me as dissonant with the rest of his discussion. He states:

> "this paper is limited to an analysis of Confucian moral evaluation of suicide, and does not touch on public policy issues. Hence the *moral* conclusions above do not have a direct bearing on *legal* issues such as the legalization of euthanasia and of physician assisted suicide, which involve issues not examined in this article, vis., the role and responsibilities of physicians, probable societal consequences (intended and unintended) of such change of law, potential for abuse, etc." (p. 95, author's emphasis).

On an initial reading, Lo's concluding comments seem rooted in a Western political tradition that is not easily reconcilable with the Chinese system of thought he so powerfully articulated. Did not Confucius himself set forth his ideals as a way for the states of his time? Were not his teachings central for the Chinese education system, which was geared toward the preparation of rulers and civil servants, who would govern according to Confucian ideals? Chichung Huang makes these points in his introduction to the *Analects*, when he notes that "[t]he Master's ethical theory is intertwined with his theory of government" (1997, p. 7). The "Way of humane government" is thus "identical with the Way of humanity that Master Kong [Confucius] inherited from the sage kings and developed into a whole system of philosophy" (p. 31). The very distinction between the moral and legal domain and the relativization of the Confucian ideals seem closer to a Western liberalism, with its account of a limited state and robust pluralism.

I think there probably is some tension between Lo's concluding remarks (rooted in a Western conceptual framework) and the Confucian tradition. However, an alternative interpretation of Lo's remarks is also possible. At the heart of the Confucian tradition, there is a distinction between the rule of virtue (the "way of humane government" characterized by the "man of *ren*") and the rule of force (the "way of inhumane government") (Huang, 1997, p. 15). Although these categories do not directly map on to the Western distinctions between morality and law, there are important similarities.

In the Confucian tradition, the Way involves the rule of sage kings who, by the power of example, motivate without force.

> "If you yourself are correct, even without issuing orders, things will get done; if you yourself are incorrect, although orders are issued, they will not be obeyed" (*Analects*, 13:6).

When "men of *ren*" come forward, their inner power brings order and does away with the need for severity.

> "If benevolent men were to rule a state a hundred years, they would be able to tame brutes and abolish capital punishment" (*Analects,* 13:11).

The rule of virtuous people contrasts markedly with the rule of those who are inhumane; who, while they may be able to maintain some power and even unite diverse kingdoms, nevertheless do not have "heaven's mandate."

The meaning given to the contrast between humane and inhumane government is at least partly determined by the historical developments that led to the ascendancy of the Confucian way in China from but one philosophical school to the establishment position for nearly two millennia (for a review of the history, see Parrinder, 1971). At the end of the Age of Philosophers, there were many warring states. One of them, the state of Ch'in, was governed by the School of Law, which advocated a harsh and repressive law in place of morality. Through military conquest, the state united China. Later, through an armed uprising, the Han dynasty overthrew the Ch'in and came into power. After several years of reflection and searching on the part of Han kings, Confucianism gradually emerged as the dominant school. Once that was done, it was natural to read the legitimacy of the Han uprising in terms of the earlier rise of the Chou (also transliterated "Zhou") over the Shang. The Chou royal house was the ideal for Confucius. Their kings exemplified the way of humanity. Thus the Ch'in School of Law, with its use of force to establish government, would be contrasted with the morality of the Han, with their espousal of the Confucian ideals (Cheng, 1981; Fields, 1983). The strength of this equation between the Han dynasty and the Confucian tradition can be found in the enduring self-identification of the Chinese as "men of Han."

This basis for the moral/legal distinction is still not sufficient to understand Lo's use, since the equation of the School of Law with the way of inhumanity and force does not allow for the more positive, legitimate meaning of law that is implicit in Lo's (and liberalism's) moral/legal distinction. However, we can identify a possible avenue for further developing the distinction. After all, the Chou revolted against the Shang, the Han against the Ch'in. Force – and by implication, law – is thus legitimate in response to the way of inhumanity (illegitimate force and law; *Mencius,* 1B:8). In this implicit philosophy of revolution, one can find an important basis for reconstructing in Confucian terms the classical Western, liberal distinction between morality and the law, as well as the notion of a limited state.

Generally, Confucianism is a very conservative philosophy, and it does not immediately lend itself to revolutionary discourse. At its heart is the notion of filial piety and the rituals, which order the full range of human interaction. The system of government is feudal, so that the filial piety central for familial relations is writ at large in relations to those who rule. At the apex of earthly rule is the sage king, whose piety is directed toward Heaven. Historically, one can read the Confucian philosophy as a moralization of previously amoral, religious practices, whereby deceased ancestors could be evoked as powers for earthly concerns, such as fertility or war. Once moralized, the *mana* (inherent power of the deceased) became the *ren* (virtue) of the humane; namely, his power of example, whereby rulership is possible without force (Parrinder, 1971, p. 306). The transnatural sanction is referred to as "heaven's mandate."

For Confucius, the Chou were only justified in their revolt against the Shang, because the latter did not have heaven's mandate. The Shang's lack of proper piety to heaven (manifest in a violation of the rituals) thus justified the absence of piety toward them. In this way, the rise of the Chou can be read as a revolt against revolution; i.e. as a use of force against those who revolt against heaven, namely, the Shang. Force is justified against force, but not otherwise. The good (=peace and prosperity) is realized by virtue, not by force.

Within the Confucian framework, there are significant difficulties in moving from this justification in principle for certain kinds of revolution to any justification in fact. There is a significant epistemological problem; namely, how does one know who has heaven's mandate? The character of that mandate is equally problematic. In order to legitimately revolt, one must know that those in power are themselves in revolt against heaven. Ideally, the sage kings will perform all the rituals correctly. However, expediency (*quan*) allows for modifications in certain contexts (*Mencius,* 4A:17), and the rituals themselves are in some need of revision as society develops (*Analects,* 9:3). One could thus not look in a simplistic way at whether every ritual was performed and conclude that if they were not, then heaven's mandate was absent. Further, the tradition itself had a markedly secular thrust that excluded any account of inspiration or revelation.

> "When Fan Chi asked about wisdom, the Master said: 'To apply yourself to the duties of man and, while revering the spirits and gods, keep away from them. This may be called wisdom indeed" (*Analects*, 6:22).

Huang (1997) celebrates this secular thrust, while simultaneously criticizing traditional Confucianism for a notion of the rituals that is "static, conservative, and at odds with the laws of evolution" (p. 10). But in this critique he does not sufficiently appreciate the epistemological problems that are raised by eliminating the concrete rituals specified at the time of the Chou dynasty. They give content to the more abstract principles of *ren* and *yi*[a]. Without the criteria they provide for judging the ruler's piety toward heaven, any revolt would be judged as the usurpation of the ruler's prerogative. The rationale for revolution thus seems to only work retrospectively, after a revolution has been successful and a new ruler is established.

However, outright revolt is not the only option within the Confucian system for responding to a government that is not fully humane. Other options are found in the doctrine of expediency (*quan*) and in suicide.

III. EXPEDIENCY AND THE CASE OF GUAN ZHONG

In a nonideal governmental system, there is a paradox that confronts a man of virtue. Such a man is bound by the rituals to have a proper respect and reverence for the ruler, but is also bound by the principles of humanity and justice to see that those principles are realized. To the degree that a ruler is inhumane, this latter commitment may require a usurpation of the ruler's prerogative. In such a crisis situation, the man of *ren* and *yi*[a] may be "expedient."

Although this doctrine is discussed at some length by Mencius (e.g., *Mencius*, 4A:17) and other later commentators on Confucius (Huang, 1997, pp. 26-28), there is only one example in the *Analects*, that of Guan Zhong. Lo discusses at some length Guan Zhong's failure to commit suicide. Two princes fought for the throne of their father, one assisted by Guan Zhong and Shao Hu, the other by Bao Shuya. Bao Shuya's prince won. Shao Hu committed suicide, when his prince was condemned. However, Guan Zhong did not, switching allegiance to the victorious prince, and assisting him in the establishment of a prosperous reign. This shift in allegiance involves a violation of the rituals. Nevertheless, Confucius seems to commend Guan Zhong for the action. The question is thus how one could justify Guan Zhong's failure to commit suicide.

Before discussing the issue of suicide, however, I would like to consider another instance where Guan Zhong deviates from the course prescribed by the rituals. In Huang's discussion of expediency, he contrasts the passages in the *Analects* that commend Guan Zhong (14:16, 17) with another that is critical of him. In 3:22, according to Huang, Confucius "censures Guan for violating the rituals in usurping the prince's prerogative" (p. 27). In that passage, Confucius states that "Guan Zhong's capacity was small indeed." He "had three households, and his house officers performed no additional duties other than their own." However, "[t]he prince of the state had a screen wall erected; Guan Shi, too, had a screen wall erected."

This passage can be interpreted as follows. The "house officers" (who were rulers of Guan Zhong's fiefs) performed their duties, and no more. They did not mark off a domain of control that was insulated from Guan Zhong's influence, and thus did not use their jurisdiction for private gain or pleasure, which would not have been in the interest of the fiefdom or Guan Zhong. However, Guan Zhong did not have the same type of restraint with respect to the prince. By erecting a barrier of privacy (the screen wall), he took a prerogative with respect to the prince that his own officers did not take with respect to him. This violation of the rituals leads to the judgment that Guan Zhong's "capacity was small indeed."

As Huang notes, this negative judgment about Guan Zhong contrasts markedly with the judgment of Guan Zhong in book 14:16:

> "Zi-lu said: 'When Duke Huan killed Prince Jiu, Shao Hu died for him, but Guan Zhong did not die.' He added: 'He was not humane, was he?'
>
> "The Master said: 'Duke Huan nine times assembled the various princes without using war chariots. It was all Guan Zhong's capability. Who can compare with him in humanity? Who can compare with him in humanity!"

Unlike Lo, Huang does not consider the difficulties surrounding Guan Zhong's failure to commit suicide, assuming that this passage is unequivocally positive. He simply asks how we can reconcile the negative appraisal in book 3 with the positive appraisal in book 14. Huang notes that "[i]t was only by applying the doctrine of expediency that the Master was able to free himself" from this paradox. Guan Zhong "was arrogant enough to have usurped the princes privileges, which was unforgivable under ordinary circumstances" (1997, p. 28). Nevertheless, because of the crisis period, and in the light of his success in bringing "peace and prosperity," Guan Zhong's activity could be overlooked.

I would like to suggest, however, that the activities of Guan Zhong discussed in 3.22 is not simply negative, and Lo (contra Huang) correctly notes that book 14 is not unequivocally positive. Further – and this will be my central point – the themes discussed in the case of Guan Zhong are all intertwined. Individual liberty and privacy (the screen wall), not committing suicide (even though a social construction of obligation, i.e. the rituals, required it), unity and harmony without force (assembling princes without war chariots), and the establishment of peace and prosperity are all intertwined. Viewed in this way, there will be important parallels between Western liberalism and the Confucian tradition.

These themes are directly related in the classical tradition of Western liberalism. After showing how they are intertwined in that tradition, I would like to return to a discussion of suicide in Confucianism, and show how East and West may not be quite as far apart as some would think.

IV. INALIENABLE RIGHTS AND WESTERN LIBERALISM

It is often assumed that the logic of Western liberalism favors the legalization of suicide, and that the opponents of such policy are anti-liberal, conservative forces. Such a contrast fails to appreciate that there are two different kinds of liberalism. In fact, the more classical liberal position, seen for example in the United States Declaration of Independence, is rooted in a notion of inalienable rights that prohibits suicide. The debate over suicide can thus be one between two forms of liberalism.

Earlier in this essay, I briefly considered what I regard as the more anemic type of liberalism; namely, one where liberty and limited government are derived from a principle of utility. Those who favor the legalization of assisted suicide and/or voluntary active euthanasia often argue that the decision about time and manner of death is a personal decision that should not be constrained by law. They thus appeal to individual autonomy. However, this argument by itself would justify much more than most want. Proposed laws usually only allow for assisted suicide in cases of terminal illness. However, if the issue is individual liberty, why constrain law in this manner? Should not any competent individual be allowed to exercise that option? Further, if another can assist, i.e. if one allows for two people to transgress the boundaries of life as long as it is by their mutual agreement, then should not such cooperative ventures be justifiable in other similar areas? For example, why not extend the right of assisted suicide to one for voluntary active euthanasia (where another is the direct agent of death, but the patient requested the action)? And if one allows for the type of cooperative activity found in euthanasia, why not extend such a right to dueling? Here you have two people, who by mutual consent, allow for themselves to be killed. This type of argument can be further extended to include war games (e.g. gang warfare), voluntary slavery, and many other activities that most people would consider inappropriate.

Although advocates of assisted suicide claim autonomy as a basis for their argument, we rarely find this argument worked out to its implications, and few consistently argue for an extension of individual liberty in other relevant areas. Thus, in practice, we find that another consideration actually plays the more important role in movements toward legalization; namely, a utilitarian assessment of the balance of pleasure over pain. Implicitly, it is assumed that the state legitimately has control over individuals, and can constrain individual action for the sake of the aggregate good of all others. The good of individuals involves a net balance of pleasure over pain, and human life is no longer worth living when suffering outweighs happiness. At that stage, life becomes meaningless for the individual (since meaning is understood in terms of capacity for positive hedon units), and it is best ended in a quick, direct manner, so that the individual and society is not burdened by the overall negative utility. In such a framework, liberty is itself instrumental, and it can be altered when it no longer leads to a maximization of overall utility. The hold of the state over the individual, manifest in the law against suicide, should thus be suspended when individual life becomes meaningless; i.e. at that stage a person should be allowed to commit suicide (e.g., when illness is terminal).

This shaky grounding for political liberalism contrasts markedly with the classical position which works with the notion of inalienable rights and limited government. John Locke is perhaps the best representative of the classical position. When Locke wrote his *Two Treatises of Government* in 1689, he was attempting to articulate a conceptual alternative to the two dominant political theories of his day, those best represented by Filmer and Hobbes.

Filmer's *Patriarcha* (1630s) provided a powerful justification of the divine right of kings and the traditional feudal system of the day, and his system has interesting similarities to Confucianism. According to Filmer, Heaven conferred authority on the first patriarch (Adam), and all political authority is derived from that first mandate. As children are bound to parents, so subjects are bound to the ruler. Individuals are thus not free and equal, any more than children are free and equal. An elder brother has priority in inheritance, and all are bound to the household patriarch. So too, political right is obtained by inheritance, and all are bound to the "patriarch" of the state. In the *Patriarcha*, property, referred to as "private dominion," is held at the discretion of the ruler, who is not accountable to the people, but directly to Heaven. What the ruler can legislate is constrained by the will of Heaven, but there is full discretion in "indifferent matters." There is thus absolute and arbitrary power in every area that is indifferent with respect to heaven.

Summarizing: for Filmer, the social and political structure is a natural one, continuous with the structure of the family. People are born unequal and unfree, and they are bound in subjection to the ruler, whose obligation is to maintain the harmony and order of society. The ruler is ultimately subject to Heaven, and accountable to no earthly authority.

In contrast, for Thomas Hobbes, men are by nature equal and free. However, through their natural liberty, which is understood as "the absence of external impediments" (1962, p. 103), different people seek to lay claim to the same things. Thereby, they become enemies of one another, and the result is a war of all against all. In order to obtain peace and preserve their lives, people thus agree to transfer their natural right to a central authority, the "sovereign," which will establish its own order. In Hobbes' *Leviathan*, the "commonwealth" arises from this transfer and centralization of power. The state is thus not natural and in continuity with family, as with Filmer. To the contrary, it is an artifact, providing a realm of order, which is in sharp contrast to the natural state of war. The commonwealth is the product of necessity; the least worst option, required to avoid a life that is "nasty, brutish and short." For Hobbes, once liberty is transferred, there is little recourse against the state and its sovereign.

Hobbes' system was in important respects progressive, drawing on the *avant garde* science and philosophy of his day. He sought to provide a systematic, empirical grounding for political theory, developing his account of the commonwealth step by step from a materialist anthropology. Filmer, by contrast, was highly traditional, attempting to provide a Judeo-Christian foundation for the feudal system. However, despite these significant differences in their conceptual foundations, Filmer and Hobbes come to similar conclusions regarding the power of the ruler with respect to political subjects. Although, in principle, Filmer's patriarch is bound to Heaven's will, and there is an obligation to promote harmony and order, in practice no subject can hold a ruler accountable. De facto, there is thus little distinction between Filmer and Hobbes. For both, the "sovereign" has absolute and arbitrary power, and revolution is never justifiable.

John Locke spoke to an age that was increasingly unsatisfied with all of the available options, but which did not yet have a sufficient conceptual alternative. The genius of Locke can be found in the way he constructively integrated elements from his predecessors, positively appropriated the scientific and religious, the progressive and traditional, and doing it in a way that led to significant limits on the state. In this way, he opened up a separate and independent domain of individual liberty and free association, and grounded a tradition of rights that would play a central role in the establishment of Western democracies, especially the United States constitutional system.

In agreement with Hobbes, Locke argued for individual freedom and responsibility in the state of nature, and he also recognized that individuals may lay claim to the same thing, and thus come into conflict with one another. However, *force* was not needed to resolve conflicting claims. According to Locke, humans are rational beings, capable of discerning moral principles, which provide a basis for resolving differences peaceably. There are thus resources within the state of nature (esp., the moral law) for the establishment of peace and harmony, and for cooperative ventures that are a part of a natural community. In this respect, he is close to Filmer.

The problem for Locke was that some individuals violate the law of nature. Instead of resolving disputes peaceably, they resort to force, transgressing the natural right of others. In doing this, they also violate their own nature as rational, and convert themselves into animals. It is this self-transgression, leading to violence against others, that results in Hobbes' state of war. But this was not the logical consequence of natural freedom and responsibility, as Hobbes contended. To the contrary, it is a violation of the natural state, preventing the realization of the peace and prosperity of natural community.

According to Locke, every individual has a natural right to defend himself and to punish any violation of the natural law. This is called the executive power, and it is used to protect natural community against those who have transformed themselves into "beasts." However, there are many practical problems, when people individually attempt to exercise this power. For example, individuals may not be strong enough, or passion may distort the way they use their force. This can result in reprisals, and an escalating state of war. In order to overcome these difficulties, we come to the basic rationale for the state.

In order to prevent war and protect individual person and property against aggression and violence, individuals consent to transfer their natural right to defend and punish (the executive power) to a central authority, which will more efficiently and effectively exercise that function. As with Hobbes, this state is derivative, an artifact created to restrain violence and war. However, this state is now distinguished from natural community, which is still possible within the limited state. Filmer's natural community becomes possible, and is the locus of peace and prosperity, when the state artifice (formed to prevent violence) appropriately performs its limited function. In this way, Locke's philosophy opened an important difference between communities of free association, on one hand, and the state, on the other. Thus, contra-Hobbes, the transfer of the executive power to a central authority does not result in absolute and arbitrary government.

Locke's state is thus constrained in two important ways. First, Locke does not look to the state for the full realization of the common good. It serves and promotes the common good by performing its appropriate function. The force and law of the state constrain violation of natural law, thus establishing the conditions for that natural activity and community, which are independent of the state and arise from free association. Here we see the implication of the classical liberal distinction between law and morality, between the state and community. With the state's function constrained by the limited purpose for which it was formed, a vast "private" domain is established for individual and communal realization of the good. Peace and prosperity are thus made possible by the state, but their realization depends on factors that are outside of the state's jurisdiction and independent of it.

However, for Locke the amount of transfer of authority from individual to state is not just constrained by the *purpose* of those making the transfer. If that were the only constraint, there would be an allowance in principle for oppressive forms of community under the umbrella of the state. For example, people could sell themselves into slavery or to a war lord. In order to prevent such inhumane forms of community, Locke will advance *intrinsic* limits on what can be transferred, determined by the natural law. In considering these intrinsic limits, we come to the key role that the prohibition against suicide plays in his political theory.

Each individual has a natural right that can in part be transferred to the state. This is the initial premise. The state's power is thus derived from the people. However, Locke will also take a step backward and ask: where did the people get their initial right? In organic notions of the state, there will be a hermeneutical circle between the question about the origin of the state's power and that of the individual's. The state's power will come from the people, and each individual's right will be conferred by the state. However, with such an organic approach, one looses an external norm by which the legitimacy of the state can be judged. In practice, the state can, in the name of some people, fail to confer rights to others, in order to realize a particular vision of community which is exclusive of central values held by others. Further, with such an organic notion, one does not have an in principle check on absolute and arbitrary government. There is also a failure to acknowledge the artificial character of the state, in distinction from natural community. Thus, Locke could not solve the question of the origin of natural rights in this way. In order to avoid absolute and arbitrary government, Locke required an intrinsic check on the scope of the state's authority. Further, his account of natural community and human flourishing independent of the state required that the state be simply derivative. It could not serve as the ground of individual rights, since its function was to protect the rights that were possessed independent of the state.

The answer to Locke's question about the derivation of individual rights is found in his account of the moral law, and its natural theological grounding. Every individual has a right to life, liberty, and the pursuit of property (or happiness; property is associated with "private dominion," which is in turn associated with the latitude to pursue human good independent of the state). This natural right accrues simply by virtue of one's humanity. The presence of natural right is a central condition of the possibility of living as a human; namely, as a free and responsible person. Basic, natural rights are such that one cannot alienate them without simultaneously alienating one's humanity and thereby transforming oneself into an animal, which the law is to protect people against. One thus cannot morally alienate certain natural rights; they are "inalienable." The source of the right is one with the source of one's humanity, not just as a given, but as a moral good. Human being and right are jointly rooted in a transcendent Ground, and there is a direct accountability of each individual to that source of reality and morality. We could refer to this as *Heaven's Mandate*, where each is directly responsible to the *will of Heaven*; namely, to the will of God.

In this way, the prohibition against suicide is at the heart of liberal theory. Since the state's power is derivative, and since no individual can alienate life or liberty, no state can claim absolute and arbitrary right over the life and liberty of its subjects. The state that exercises such an absolute right is by definition inhumane. In the words of Locke:

> "Though the legislative ... be the *supreme* power in every commonwealth; yet, ... [i]t is *not*, nor can possibly be absolutely *arbitrary* over the lives and fortunes of the people: for it being but the joint power of every member of the society given up to that person, or assembly, which is legislator; it can be no more than those persons had in a state of nature before they entered into society, and gave up to the

community: for no body can transfer to another more power than he has in himself; and no body has an absolute arbitrary power over himself, or over any other, to destroy his own life, or take away the life or property of another." (1980, sec. 135)

Summarizing, in the classical liberal tradition, the issue of suicide cannot be isolated and placed within the realm of private decision-making. To the contrary, the prohibition of suicide is central for the very fabric that sustains the private domain. The distinction between law and morality (and the related distinction between state and church, with concomitant notions of religious liberty), as well as the liberal account of individual rights and a limited state, are all systematically intertwined with the respect for life found in the notion of an inalienable right.

When Lo contrasts Confucian and Western views of suicide, he does not really convey the force of this classical Western tradition. He highlights the discussions surrounding anemic liberalism and its attempt to privatize the self-regarding suicides of the terminally ill. That form of liberalism will provide little inspiration for China, and the case examples will compare poorly with the courageous and inspiring examples in the Confucian tradition. Missing from Lo's contrast is a sufficient formulation of the Western rationale for a prohibition against suicide, a rationale that is intimately intertwined with the whole classical liberal tradition of individual rights and limited government. Here the exemplary case involves the individual who is genuinely tempted to make an exit, but who nevertheless resists that temptation out of respect for higher moral principles and the fabric that sustains a peaceful and prosperous society. The inspiring example in the West involves the individual who rests assured of his or her humanity in the midst of suffering, human limitation, and death, thereby remaining human and responsible in the face of dehumanizing forces.

This noble example is not just found in the West, however. As we saw at the end of the previous section, the same factors that are intertwined in classical liberalism are also found in the case of Guan Zhong.

V. THE WISDOM OF GUAN ZHONG

The prominent options in current Chinese political theory have interesting similarities to those that John Locke encountered. As with Filmer's *Patriarcha*, the Confucian tradition looks back to a feudal, patriarchally structured society, in which political subjects are bound to rulers by loyalty that is but the extension of the fabric of family. Similarly, the current Marxist philosophies, with their claim that power rests in people who are equal, has interesting similarities to the thought of Hobbes. Although the notion of government is organic rather than atomistic (as with Hobbes), it is justified by a similar claim to embody the "scientific" spirit of the day. The systems of Hobbes and Marx are both materialist, and they point to an antecedent period of strife and suffering (with Hobbes it is a state of war, with the communist regime it is a class conflict). The language of "evolution" and progress is used to distinguish "forward looking" egalitarian thought from the "backward looking" feudal thought.

It is natural for Chinese philosophers to read this contrast between Communist and Confucian thought in terms of the contrast between Ch'in legalism and Han Confucianism. On one side, those sympathetic to Chinese Marxism argue for the Legalists, and align their thought with the rule of law, equality, and a public-minded humanism (Fields, 1983). On the other side, critics of current Chinese government seek to recover Confucian forms of humanism, arguing that "[t]he Legalist government may succeed in ordering society, but the consequent social order is imposed from outside, not ingrown within, and therefore will not last and will in fact collapse because it is against the inner wishes of the people." (Cheng, 1981, p. 295). They point to a social fabric that is independent of the state's mechanisms of force, and emphasize virtue and the power of example.

The Confucian tradition captures the heart and history of China, and seems central for sustaining the fabric of society. The communist ideas make claim to the head of China, advocating the importance of science and technology, and championing the workers who seemed excluded in the Confucian feudal system. However, as with Filmer and Hobbes, these diverse traditions seem to arrive at a similar result; namely, an unlimited and arbitrary government, which de facto subjects the individual to the collective. In the case of Communism, the individual is subject to the state's tyranny; in the case of Confucianism, it is to the tyranny of society, with its rituals. Neither option seems adequate, leading to the contention by some that there is a crisis in current Chinese political theory.

Interestingly, within this contrast, Western liberalism does not appear as a separate option. Either it is aligned with Confucian thought, as if they were sister traditions in their resistance to statism, or liberalism is associated with its anemic, utilitarian form. In this latter case, liberalism is rejected by both sides, since its emphasis on private, self-focused pleasure contrasts with a public-mindedness that is central to the Chinese mind set. For some Chinese philosophers, Western liberal thought simply means the rationalization of financial gain (and thus the market). With this liberal emphasis comes an indulgent, pleasure seeking, pain shunning focus, which is the epitome of vice, rather than virtue. The cases of self-regarding suicide that Lo highlights fit well with this anemic liberalism.

Lo's essay does with the Confucian tradition what Locke did with the tradition embodied in the thought of Filmer. Whether or not he intended this, his account presents the Confucian tradition as a viable alternative, not simply as an outdated or benign system of historical interest. In order to do this, however, the tradition must be modified, so that it is responsive to the needs of today. Certain elements must thus be emphasized at the expense of others. In his analysis of suicide, Lo will abstract certain principles (two theses and two antitheses), and he will give priority to notions of humanity (*ren*) and justice (*yi*[a]). He chooses to downplay the aspects of the Confucian tradition that are tied to its feudalism or to notions of shame. However, there is tension in his analysis, one that can be found in his interpretation of Guan Zhong.

Lo opens his essay with a discussion of the Confucian view of human life. Quoting Confucius and Mencius, he shows that life is not an absolute value. It may need to be sacrificed for the sake of the higher values of humanity and justice. This view is formulated as his first main thesis:

> "Confucian Thesis I: one should give up one's life, if necessary, either passively or actively, for the sake of upholding the cardinal moral values of *ren* and *yi*[a] " (p. 72).

Near the end of his essay, Lo considers the case of Guan Zhong, and asks how his failure to commit suicide can be justified. Noting that there was controversy over this case, and many within the Confucian tradition sought to explain away Confucius' positive appraisal of Guan Zhong, Lo points to the "important breakthrough" in interpretation that came during the seventeenth century. When the Manchus defeated China, many committed suicide, expressing their dedication to the perishing Ming Dynasty. However, three Confucian commentators criticized mass suicide, arguing that it involved an inappropriate understanding of duty. Lo notes:

> "these three Confucians did not reject dying for integrity altogether. They only intended to modify Confucian Thesis I by broadening the object of dedication from an emperor, a royal house, and a dynasty to the entire country. They resolved to stay alive as long as they could still contribute to China in some way ... " (p. 90)

In the light of this commentary, Lo formulates his Antithesis I.

> "Antithesis I: One should broaden the scope of one's commitment; instead of dying for a rather limited cause, one should live and die for an object of a higher order." (p. 88)

It is clear from his discussion that he considers "China" to be the "object of higher order."

There is a curious dissimilarity between Lo's thesis and antithesis. In the thesis, *ren* and *yi*[a] are the principles for which one must die. Accountability is ultimately to Heaven. However, in the antithesis, one lives or dies for a cause, and his interpretation makes clear that cause is China. Why is China put forward as the "object of a higher order"? This nationalism does not seem to be fully consistent with Lo's implicit revision of the Confucian tradition, where he points to general principles of humanity and justice, rather than to a particular emperor, dynasty, people, or even nation. How does one reconcile these different approaches to accountability?

Put simply, one can say that while the moral accountability is ultimately to Heaven, and to the principles of *ren* and *yi*[a], these moral principles are only given content in particular contexts, where the obligation is to a contingently given set of norms (the rituals) and to a particular people. Chung-ying Cheng clarifies the tension (1998, pp. 142-143):

> "It is important to recognize that nowhere in Chinese history or philosophy was a ruler considered, or would consider himself, to have his primary obligation to Heaven and not to the people. Heaven is only the de jure basis for political rule; the de facto foundation of rule, among Confucians, is always the people."

Here, the notion of "rule" can be extended to refer not just to those formally in government, but to any person of virtue. For all people, the focus of obligation should be on "the people."

The emphasis placed upon this de facto social accountability explains moral limits on self-regarding suicide in the Confucian tradition. As Lo notes (pp. 78-79), filial piety (which is also the basis for obligation to broader social units) constrains what a person may do with self.

> "[C]hildren are permanently indebted to parents because children do not exist on their own, but owe their existence to the parents. If one is not the author of one's biological life, how can one have the autonomy to dispose it as one wishes? Suicide is then understood as usurping the authority of parents. In short, unless filial piety is outweighed by another moral value such as *ren* and *yi*, the former is usually a moral reason strong enough to forbid suicide."

As has already been noted, in practice it will not be abstract principles of humanity and justice, but rather de facto obligations to a broader social unit (China) that may outweigh the particular obligation to a parent, and thus require the suicide. The fact that it is for "the people" makes it other-regarding.

The other-regarding emphasis (tied to the de facto obligation to the people) also explains Lo's Antithesis I; namely, the need to "broaden the scope of commitment; instead of dying for a rather limited cause, one should die for an object of a higher order." When a higher cause demands it, one should refrain from committing suicide, even if the rituals demand it.

Earlier, I quoted Confucius' commendation of Guan Zhong, especially his capacity to assemble the princes without war chariots. There is another similar passage in the same book of the *Analects*. This nicely illustrates Ping-Cheung Lo's Antithesis I.

> "Zi-gong said: 'Guan Zhong was not a man of humanity, was he? When Duke Huan killed Prince Jiu, he not only was unable to die but became the duke's prime minister, instead.'
>
> The Master said: 'Guan Zhong helped Duke Huan become overlord of the various princes and set everything right in the empire. The people to this day benefit from his favors. But for Guan Zhong, we would be wearing our hair loose with our garments fastened on the left. How could we expect him to be obstinately truthful like a common man or a common woman and hang himself in a gully without anyone knowing about it?'" (14:17)

The historical defenders of Guan Zhong point to his capacity to promote prosperity, and to preserve the culture of Chinese civilization (i.e. prevent "wearing our hair loose with our garments fastened on the left," the customs of barbarians). Because of these abilities, Zhong is distinguished from the "common man or the common woman." In other words, his life is instrumentally regarded as a means to the good of China. His elite status and unique capacities exempt him from an obligation that would fall on those of lower status. On the other hand, the fact that Guan Zhong set up a screen wall – that he opened up a private space, where he was not transparent in his accountability to the prince and to the people – showed a deviance from the "higher cause," leading to the negative judgment upon him that is given in 3:22.

It is here, in the Confucian appraisal of Guan Zhong, that we can see the difference between the Confucian tradition and the Western liberal tradition associated with inalienable rights. In the Christian tradition that influenced Locke, it is not the parents that ultimately give life. It is God. Unless God were to directly command the taking of life, it would be presumptive for anyone to do so, whether of self or others. There is thus a modification that goes beyond the one made to the Confucian tradition with the fall of the Ming Dynasty. The "object of higher order" is not just emperor, dynasty, or even China. It is Heaven itself. One thus loses the distinction between the de jure and de facto foundation of political rule. With this final move in universalizing the cause, there is qualitative change in the notion of suicide. Lo's Antithesis I, which is a provisional constraint on suicide, becomes generalized. The result is a Lockean inalienable right; namely, a right to life (the mandate) that cannot be alienated. By making each individual immediately accountable to Heaven, the elitist, feudal framework is fundamentally altered. No longer is the "mandate of Heaven" passed on through nation, dynasty, and emperor. A space is opened up between state and society, on one side, and individual, on the other. The individual is accountable to Heaven, and to this extent, is above the state. It is from this notion that individual liberty is generated; namely, the recognition of a private space, which is a condition for virtue, and which may warrant action that is above social norms and thus viewed as transgressing those norms.

The Western liberal view implies that the prohibition of suicide and the establishment of a domain of individual liberty and privacy are linked in the notion of an inalienable right. As a result of this link, there is an alteration in the appraisal of Guan Zhong. The establishment of the "screen wall," which was negatively appraised in the Confucian tradition, is now seen as the corollary of Guan Zhong's violation of the ritual demanding suicide. When Guan Zhong erected a screen wall, he confronted the authority of the king/state by opening up a space for his own activity. In this way, he risked his own life. His refusal to commit suicide can thus not be read as a cowardice. The refusal to commit suicide (to actively take his own life) must be read together with his willingness to risk his life (and thus not give priority to life). Taken together, they highlight a nonviolent revolution against the inhumanity of a state and society that oversteps its bounds. Through such "expediency," Guan Zhong makes the way for the exercise of virtue, and his accountability is directly to heaven, from which he receives his mandate.

At the heart of this interpretation is the distinction between law and force, on one side, and morality and virtue, on the other; a distinction that is well established in the Confucian juxtaposition of the ways of inhumane and humane government. Prosperity and the exercise of virtue go together, and both depend on the state's recognition of its own limits. Guan Zhong used the space he opened up to realize human good. The king made this possible by accepting Guan Zhong's space; in other words, by not violently intervening in the face of Guan Zhong's resistance to the state's inhumanity. The ensuing prosperity thus depended on two things: (1) the courage of Guan Zhong to open up a space independent of the state, where he could exercise virtue, and (2) the state's (Duke Huan's) willingness to cede space, thus accepting its own limitations. The passages in the *Analects* that commend Guan Zhong can also be understood as commendations of Duke Huan, who recognized the capacity of Guan Zhong, and thus did not prevent his construction of a private space (the screen wall).

In this interpretation of Guan Zhong, the prohibition of suicide (it is aligned with the inhumanity of one who is base) serves a function similar to an inalienable right in Western liberalism. It simultaneously places a limit upon the activity of an individual and a state, sustaining the legal/moral distinction that provides the condition of a natural peace and prosperity. Within the Confucian tradition one thus finds certain core motifs of Western, classical liberal thought exemplified, although the appraisal provided within the tradition has not yet unfolded that germinal wisdom. Classical liberal thought associated with Locke can be viewed as a celebration of the dignity and wisdom of Guan Zhong.

Department of Philosophy and Center for Bioethics
University of South Carolina
Columbia, South Carolina, USA

REFERENCES

Cheng, C.-Y.: 1981, 'Legalism versus Confucianism: A philosophical sppraisal,' *Journal of Chinese Philosophy*, 8: 271-302.

Cheng, C.-Y.: 1998, 'Transforming Confucian virtues into human rights: A study of human agency and potency in Confucian ethics,' in W.T. de Bary and Tu Weiming (eds.), *Confucianism and Human Rights*, Columbia University Press, New York, pp. 142-153.

Durkheim, E.: 1951, *Suicide: A Study in Sociology*, J. A. Spaulding and G. Simpson (eds.), Free Press, New York.

Fields, L.B.: 1983, 'The Legalists and the fall of the Ch'in: Humanism and tyranny,' *Journal of Asian History,* 17: 1-39.

Filmer, R.: 1949, *Patriarcha and the Other Political Works of Sir Robert Filmer*, Oxford University Press, Oxford.

Hobbes, T.:1962, *Leviathan*, M. Oakeshoot (ed.), Collier Books, New York.

Huang, C. (tr.):1997, *Analects of Confucius*, Oxford University Press, New York and Oxford.

Lau, D.C. (tr.): 1984, *Mencius,* Chinese University Press, Hong Kong,

Lo, P.-C.: 1999, 'Confucian views on Suicide,' in this volume, pp. 69-101.

Locke, J.: 1980, *Second Treatise of Government*, C.B. Macpherson (ed.), Hackett Publishing Company, Indianapolis, Indiana.

Parrinder, G. (ed.): 1971, *World Religions: From Ancient History to the Present*, Hamlyn Publishing Group Limited, New York, Ch. 17.

Quill v. Vacco, 1996, 80F. 3d 717.

EDWIN HUI

A CONFUCIAN ETHIC OF MEDICAL FUTILITY

I. INTRODUCTION

The concept of medical futility is a complex one and has received renewed attention in the past few years,[1] particularly in association with treatments such as advanced life support, CPR, and the provision of nutrition and fluids, for patients who are either terminally ill or in a so-called, for lack of a more humane term, "permanent vegetative state." While we realize that a consensus is far from being reached within the Western medical community, the purpose of our attempt in this chapter is to try to articulate a Confucian perspective on this issue with the hope that such a perspective, from a cultural viewpoint considerably different from the West, may contribute some useful insights to the discussion. Also, futile treatments are often a result of modern medical technologies whose enormous power to extend life is largely unknown to most of the developing Asian countries. Since these countries are rapidly implementing programs of modernization that include the importation of advanced medical technologies, a Confucian notion of medical futility may prove to be quite necessary as people in China and her neighbors who are predominantly under the influence of the Confucian tradition begin to encounter the implications of "high-tech" medicine in their cultural milieu.

Furthermore, in recent years, North America has witnessed a massive immigration from Asian countries, particularly Chinese from Hong Kong, Taiwan, southeast Asia, and, increasingly, the People's Republic of China. It has been noted that even when a person has been significantly acculturated into the foreign country to which he or she has migrated, at times of life crisis (such as major illness or death) ideas, values, and worldviews adopted earlier in life tend to resurface and structure one's response (Barker, p. 251), therefore playing a big part in determining the patient's response to care. For this reason, even though cultural differences manifest themselves in different forms, e.g., lifestyle, education, diet, religion, etc., it is in the area of health care and specifically in the treatment of terminal patients that we see the most difficult and sometimes

Ruiping Fan (ed.), Confucian Bioethics, 127–163.

painful conflicts that can result in confrontations between health care providers of the West and health care recipients of the East. A Chinese Confucian ethic of medical futility may serve to clarify some of the cross-cultural issues surrounding the decision making process in the actual medical practice of caring for Chinese immigrants who are terminally ill.

II. MEDICAL FUTILITY: DEFINED

A futile medical procedure has been defined as one that cannot achieve its stated goals or produce its expected benefits with an acceptable level of probability regardless of repetition and duration of treatment (e.g., Hansen-Flasche, p.1192). But in practice, because goals and benefits as well as estimations of probability are dependent on both scientific and nonscientific factors that may affect perceptions of the illness and the predictions of outcome, the notion of futility is a highly value-laden and culturally-bound issue. For this reason, to date no objective unambiguous criteria have emerged in the determination of the futility of any medical treatment, and most writers agree that value choices are involved in most futility judgments (See Schneiderman, et al.; Veatch and Spicer; Schneiderman and Jecker). Precisely because this concept is not value-neutral, it brings cultural differences into sharp focus and serves as a useful tool for exploring the cross-cultural dimensions of medical ethics. As a corollary, even if a set of criteria for the determination of medical futility can be established, the communication of that information to patients of different cultures is itself an ethical issue, adding to the complexity and richness of the cross-cultural aspects of its ethical analysis.

III. MEDICAL FUTILITY: DETERMINED

Futility judgments are complex and difficult in actual medical practice because there are no universally accepted clinical criteria to declare a particular intervention futile (Lantos, et al., 1988, p. 82). Both qualitative and quantitative approaches have been proposed (See Schneiderman, et al., pp. 949-954), but agreements are hard to reach in (A) establishing the goals of a certain medical intervention (qualitative approach), and in (B) allowing the probability of success/failure to determine futility (quantitative approach). We will review some of these approaches.

(A) Goals of Treatment as a Determinant of Medical Necessity or Futility

A goal of a particular therapy may be viewed in at least three ways: (i) in strictly physiological terms, (ii) in terms of the length of life, and (iii) in relation to the quality of life. A treatment could be judged futile if it does not achieve one or more of these goals (Youngner, 1988, pp. 2094-5).

(i) Physiologic Terms

"Physiologic futility" can defined as the failure of a particular medical procedure to achieve certain objectively demonstrable physiologic effect(s), and, in general, we may assume that physicians are qualified and knowledgeable in medical science to make such a determination. Some physiologically futile examples are quite unequivocal, as in applying cardiopulmonary resuscitation to a patient who is known to have last respired several hours ago (Veatch and Spicer, p. 18), or treating a patient with biopsy-proven stomach cancer with interferon when it has no known effect on stomach cancer (Miles, 1994, p. 233). But, not all physiologically futile determinations are as straightforward. For example, in some cancer chemotherapy trials, a fifty percent reduction in tumor size is considered to be therapeutically effective, while any lesser response is considered to be physiologically futile (Faber-Langendoen, p. 832). Here, one cannot assume that the judgment is an entirely objective, "value-neutral" and scientifically/physiologic determination, for the design of the trial, the choice of the statistical threshold, the selection of data, the method of collecting the data, and the language and manner of reporting and describing the data are inescapably value-laden, influenced by the physician's values, beliefs, and worldviews. Increasingly, western philosophers and scientists have begun to recognize that even apparently empirical and objective fields of knowledge, including medical sciences, are inescapably value-laden, involving personal value judgments and belief systems in the processes of forming hypotheses, measuring outcomes, and drawing conclusions (See Polanyi, 1962). This is not to say that physicians and medical scientists are therefore not entitled to make judgments on 'physiologic futility'. Rather, it emphasizes that there are nonscientific components inherent in these and other so-called scientific/medical judgments that must be recognized and, when appropriate, disclosed and communicated to patients. In the above example, a patient may conceivably value the lesser (<50%) response differently; such a response may satisfy the emotional needs or other symbolic goals of the patient or his family, and may in turn lead them to consider such a response medically effective rather than futile.

However, in this case the patient, the physician/scientist, and the family must recognize that their respective judgments of physiological effectiveness may have been influenced by a rather disparate set of values, needs, and goals. And in the event of a disagreement, it should be acknowledged that the disagreement is not entirely based on scientific facts, but human values as well. Or to put it slightly differently, rather than treating scientific/medical data as the decisive factor in determining physiologic futility, it is seen merely as the occasion for the expression of divergent or conflicting values. Any attempt to reconcile a disagreement may need to take the respective values of all parties into consideration.

(ii) Length of Life Considerations

Treatment goals in relation to length of life are also points of uncertainty in futility judgments because different parties may bring divergent values to bear on the situation. A specific example is provided in the scenario of a terminal cancer patient with widespread metastasis who has experienced a cardiac arrest where resuscitation would only restore cardiopulmonary functions temporarily. If it is clinically determined that impending death is inevitable with no chance for patient survival to discharge, and further aggressive treatment in fact interferes with the appropriate care of the dying and inflicts unnecessary suffering, such a short-lived and transient benefit is often considered not to be a worthwhile goal and, hence, futile (See Bedell, *et al.*, pp. 569-76). But a patient or family with different values and worldviews may deem such intervention to be of benefit, or even of necessity, because they may value any life span gained, however brief, to be worth the effort. In this case, the conflict that has to be resolved between the patient and the physician(s) is again one of value over the limited benefit the procedure CPR brings, and not over the effectiveness of the procedure. Both parties may agree that the intervention is of limited effectiveness (in the sense that the effect is not sustainable), but the value of the limited benefit may be perceived differently by the two parties involved. It is not justifiable for physicians to declare unilaterally and arbitrarily that such interventions are medically futile. A more reasonable approach would be to expose the differences of value and accept them as a starting point in any attempt to reach a resolution.

(iii) Quality of Life Considerations

Difficulties in determining treatment goals relative to quality of life are best illustrated by clinical cases in which the patient has no chance of regaining neocortical function, as patients in a "permanent vegetative state." Because Schneiderman *et al.* believe that "the goal of medical treatment is not merely to cause an effect on some portion of the patient's anatomy, physiology, or chemistry, but to benefit the patient as a whole," and because these patients are unable to substantively recognize or appreciate any benefit, physiologic or emotional, and their state of existence is often considered "better off dead", all medical treatments, including provisions of artificial nutrition and hydration, would predictably yield "qualitatively poor results" and are to be considered futile, and need not be offered to patients as options (p. 950). Furthermore, these same writers claim that in a situation in which survival requires the patient's complete dependence on intensive medical treatment, and renders the patient incapable of achieving any other life goals (thus obviating the goal of medical care), medical treatments that may be effective are not considered to be beneficial, they are deemed futile. And not only do they not need to be offered to the patient but the patient's family has no right to demand them (p. 953).

But, as Lantos *et al.* point out, depending on one's worldview, values and belief systems, what is disvalued may be considered valuable, and feeding or providing other treatments to patients in a "persistent vegetative state" may serve goals treasured by the patient, family, or society (Lantos, *et al.*, 1989, pp. 81-84). These goals may not be irrelevant to futility determinations and automatically excluded just because the patient cannot return to cognitive life. It is risky for the medical profession to have to decide on another person's goal, quality and value of life, even when what is being considered is merely vegetative life. As some have commented, "the physician in no way could claim expertise in knowing the value of their patient's vegetative life" (Veatch and Spicer, p. 21). Quality of life, despite recent efforts to quantify it, remains a subjective and relative notion and defies any attempts of objectification (Crisp, pp. 171-183).

(B) Quantitative Prediction:
Probability of Success as a Determinant of Futility

In view of the value-laden nature of treatment goals, Schneiderman et al. have proposed a quantitative approach to measure the efficacy of medical treatments which may offer a more objective means to determine futility

(Schneiderman, *et al.*, pp. 949-954). What they propose is that if a medical treatment intervention has been useless or unsuccessful in the last one hundred cases, it should be judged futile. While this appears to be quite reasonable on the surface, several questions need to be answered before one applies this quantitative definition. Statistically, this formula amounts to justifying a determination of futility when the physician is ninety-five percent confident that a procedure would be successful no more than three times out of one hundred. But how does a physician decide on this particular statistical cutoff point, and not the other? This medical decision must somehow be influenced by the decision-maker.

Secondly, to decide that the procedure has indeed not been useful or successful in the last one hundred cases, or whatever chosen number of cases, requires an evaluation based on data either from "controlled clinical trials" or the physician's "extended experience" (Schneiderman, *et al.*, pp. 949-954). Clinical trials as a source of statistical prediction must be used with caution because of their limitations (due to, for instance, heterogeneity and size of patients) and susceptibility to errors (social and psychological factors) (See Lantos, *et al.*; Truog, *et al.*). Physicians' experiences, while undoubtedly valuable in most clinical situations, are subject to personal bias in their memory recall and application.

Also, Waisel and Truog rightly point out that quantitative assessment is often influenced by the qualitative assessment of the same issue. Just as the assessment of the odds (quantitative) of purchasing a lottery ticket is influenced by the lure of the lottery prize (qualitative), similarly the assessment of the success of a medical procedure is influenced by both the "quantitative likelihood of success as well as the qualitative aspect of success." We are inclined to agree with their conclusion that "quantitative and qualitative futility are intertwined, and occasionally, the determination of quantitative futility becomes a value judgment or a qualitative assessment of futility" (Waisel and Troug, p. 308).

Some writers also point out the narrowness of a quantitative definition, reducing the usefulness of a medical procedure to a "purely medical effect" when it could be significant in other ways, as in terms of family well-being, extra time with family and friends if it is deemed valuable to the patient, and it may even be helpful for members of the medical team (Lowey and Carlson, p. 429). These authors therefore suggest that quantitative determinations are better made jointly by physicians and other community partners. One final problem is the distance between medical statistics in futility determinations and clinical decisions to deny a particular patient treatment based on statistical futility. Both theoretically and in practice, it is a recognized fact that statistical inferences about what might happen to a population of patients do not allow one to claim certainty and accuracy in predicting what will happen to the next particular patient. If, in fact, "the term 'medical futility' should be applied only to a specific intervention, for a specific patient, at a specific point in time" (Jecker, p. 141), quantitative assessments must be applied with caution, taking the qualitative and evaluative aspects of futility judgment into serious consideration.

These various considerations suggest that determinations of futility, regardless of the approach one adopts, be it physiological, qualitative, or statistical, cannot be considered value-free, morally-neutral guides for clinical policy and medical decision making. Many have begun to see that values, be they cultural, social, religious or personal, influence not only what is identified as scientific/medical facts but also the selection of these facts for consideration. Moreover, the manner data is used and interpreted is often a direct function of the values brought to bear on those facts by its users (Hansen-Flasche, p. 1192). In this way, one must see value judgments as an inherent part of any treatment decision, including determinations of futility.

In this regard, there is apparently agreement between both "proponents" and opponents of the concept of medical futility. Jecker has said that it is not reasonable to expect health care decisions "to be purged of ethical and other value components. Instead, it is generally recognized that every health care decision incorporates value components" (Jecker, p. 140). Veatch and Spicer also note that "there simply is no such thing as a value-free and concept-free fact" (p. 19).

IV. VALUE CONFLICTS IN FUTILITY JUDGMENTS

When one acknowledges the essential role of values in futility judgments, one immediately faces the problem of the plurality and possible conflict of values between physician and patient, especially in a cross-cultural context. Should the decision made be based on the physician's or patient's values, or both? In this regard there exist three views: (i) decisions limiting treatment should be based solely on the patient's values as part of patients' right to autonomy (See Younger, 1990, p. 1295); ii) futility judgments are part of the authority and responsibility of the physician and other medical professionals and are not required to take patient or family values, beliefs, and consent into consideration (See Schneiderman et al. pp. 949-954; and Task Force, pp. 1435-39); and (iii) a balanced approach with a strong patient or surrogate role in decision making in concert with physician participation (See American Thoracic Society, pp. 726-731; Tomlinson and Brody, p. 1280).

(A) Priority of Patient Values

Today, both medical ethics and case law in many western societies give primacy to patient autonomy – the right of patients to be fully informed participants in the decisions surrounding their medical care (Task Force, p. 1437). This means that patient values are explicitly given priority in the medical decision making process. The assignment of priority to patients' autonomy in the hierarchy of medical decision making is echoed in Angell's comments on the Helga Wanglie case in which she stated that the sources of the decision to refuse medical treatment are "... in order of preference, the patient, the patient's prior directives or designated proxy, and the patient's family" (p. 511). Specifically, with regard to futility judgments, Lantos *et al.* argue that patient preferences are an essential component of that decision and that the patient may insist on therapy that the physician believes will not be beneficial: "When the chance of success is low, but the alternative to treatment is death, and patient desires therapy, the presumption should be in favor of treatment" (1989, p. 83). This conclusion is also reflected in the US President's Commission on Ethical Problems which recommends that the patient's and/or surrogate's wish and approval must be sought in the assessment of the futility of resuscitation and that ultimately the physician should follow the patient's or surrogate's decision.

In a culture where individual autonomy is held as sacrosanct, many are of the opinion that patients' role in futility determinations include the right to have their values and beliefs taken into consideration, the right to have medical and scientific complexities explained exhaustively and compassionately to ensure that informed choices can be made, and for physicians' deliberations of futility decisions to be undertaken from the patient's perspective and in the patient's best interests, which often include the patient's value systems and worldviews.

(B) Priority of Physician Authority

While a high level of respect for patient autonomy in the medical decision making process is generally commendable, many believe that issues of futility judgment may provide the occasion to reflect on the limit of such an autonomy. Some point out that futility judgments are precisely when patient autonomy should begin to yield to physician authority (Hammond and Ward, pp. 136-138). For this reason while paternalism, which has long characterized the physician-patient relationship in medical practice, has given way to a general respect of patient autonomy, in futility judgment, there persists a tendency to allow the physician's judgment to overrule the patient's or surrogate's autonomous choice. This has often been urged in CPR cases with adults (See Tomlinson and Brody; Blackhall; and Murphy), for CPR in low-birth weight neonates (Lantos, *et al*., 1988, pp. 91-95), for fluid resuscitation for severely burned patients (See Hammond and Ward), and extracorporeal membrane oxygenation for the terminally ill (See Troug, *et al*.). Likewise, the American Medical Association guidelines on resuscitation, while acknowledging and supporting patients' judgment on the quality of life, nevertheless grant to the physician the right to judge futility and to limit treatment on that basis (1991 Council Report, pp. 1868-71).

Several arguments have been advanced to grant physicians such full authority in futility determinations. Some feel that only physicians are equipped with the unique scientific knowledge and clinical experience to be qualified to make decisions concerning treatment futility for patients (Murphy, p. 2099). Also, it is a professional duty for physicians to maintain certain professional standards of practice upon which futility determinations are often based. Others have contended that futility decisions are largely physicians' domain because, as members of a moral profession, physicians have a moral obligation and a socially sanctioned responsibility to evaluate, promote, and protect patients' best interests, and consequently to withhold therapies that are not beneficial to their patients. In other words, "this authority flows from the social concern with their professional rather than their personal or individual integrity" (Johnson, p. 1366; see also Tomlinson and Brody, p. 1279).

Respect for physician autonomy has also been advanced as another reason that futility decisions should be a professional judgment which takes precedence over patient preferences, allowing physicians to make decisions without being subject to patient approval (Schneiderman, et al.,

p. 953). In distinguishing medical effect and benefit, these writers argue that physicians are not obligated to provide non-medical benefits such as keeping a patient alive in a persistent vegetative state because such an action provides no medical benefit. In such an exercise of physician autonomy, it is claimed that "physicians and other care givers have a legitimate interests in seeing that their knowledge and skills are used wisely and effectively, and to be pressured to perform interventions that are not believed to be appropriate removes the dignity and sense of purpose in the practice of medicine" (Troug, *et al.*, p. 1562). Lastly, it has also been argued that futility judgments by physicians can be endorsed because ultimately they protect and promote the autonomous choices of patients or surrogates because "it is inherently and unavoidably misleading to offer a futile treatment, and so it is corrosive of [a patient's] autonomous choice to do so" (Tomlinson and Brody, p. 1279). In other words, physicians can frustrate patient autonomy by offering or allowing patients to choose treatments that are considered futile, and thus confuse the issue by implying a benefit and a real choice when none really exists (Tomlinson and Brody, p. 1279; see also Younger, 1990, p. 1295). This hinders the patient's ability to make a well-informed, autonomous decision.

Quite clearly all these considerations which give priority to the physician in deciding futility involve many aspects of a physician's professional life: qualifications, experience, duty, obligation, dignity, and autonomy. In all these aspects, physicians' values and personal convictions are involved and impact their judgment in making a futility determination. To give physicians priority in futility determinations runs the risk of a physician's values and interests being protected at the expense of a patient's values and interests, and may lead to the possibility of abuse. But the fact that physicians' values are involved in health care decisions does not mean that those values should be considered irrelevant or that physicians should have no part to play in making those decisions. Rather, these considerations demonstrate that futility determinations are highly complex medical decisions that should demand the respect of the values of all the parties concerned those of the patients and/or the patient's family, as well as the physicians and the other health care team members. There is a reasonably strong feeling in the medical community of the West that in order to preserve 'medical futility' as a viable and clinically useful concept, values and convictions inherent in the medical profession as a whole, though not necessarily any particular individual member, should not be lightly dismissed.

(C) Moving Toward Shared Decision Making

Because futility determinations, like other clinical judgments, combine technical considerations, patient and physician values, and clinical data, it is suggested that the framework for these determinations should be one of shared decision making (See American Thoracic Society, pp. 726-731). A model of shared decision involves the patient and physician jointly determining the goals of treatment, its likely outcome, the potential benefit it may bring, and the criteria for determining failure or futility. The patient and his/her physician(s) may not agree, but in the process their respective values, preferences, and prejudices must be frankly revealed, discussed, and mutually respected. This way, each member of the relationship is empowered to bring the relevant expertise, facts, and values to bear on the decision making process. Also, given the asymmetry of power, status, and knowledge between physician and patient, a physician should consider it his or her professional duty to initiate, facilitate, and sustain such a joint discussion. When repeated discussions do not result in an agreed course of action, one of the following may take place: (a) The physician may appropriately maintain his/her professional determination of futility, but make exceptions out of respect for the patient's special needs, values, and beliefs, "provided such exception do not impose undue burdens on other patients ... by directly threatening the health care of others" (Schneiderman, *et al.*, p. 953); (b) However, when the patient's insistence on such care begins to impose undue burdens on the physicians and the institution, the limit of patient autonomy will have to be assessed and addressed in a public forum. Most people do not believe that physicians should be given the power to make a unilateral decision to withhold or withdraw treatment professionally deemed "futile" when that opinion is not shared by the patient. The public forum may take place in a variety of venues such as hospital ethics committees, arbitration boards, or even the courts. This is not to say that socially sanctioned standards always override patients' decisions. This would violate the patient-oriented conception of medical care and move towards a social conception which runs the risk of subsuming patient and physician values under a broad socially-sanctioned standard. Rather, this move is sometimes necessary because in certain critical clinical situations, the resolution of conflict may require "an explicit public process of social decision making" (Truog, *et al.*, p. 1563). Also, such a social process is itself an expression of shared decision making. Public forums allow patients as a group to engage in public discussion on facts and values of both patients and physicians which bear on judgments of futility. The best safeguard against both professional and patient arbitrariness is not more arbitrariness, "but rather an effective social dialogue which can ensure that the value judgments ... have an adequate social warrant" (Tomlinson and Brody, p. 1280). In a cross-cultural context, the shared model is preferred since cases are treated on an individual basis, involving the determination of the case-specific values, goals, and benefits of the patient, the patient's family, the physician, and other health care providers involved in the treatment decision.

Given that patients' and providers' values are recognized in the shared model of futility determinations, in the rest of the chapter we will discuss a specific value system, namely the Confucian tradition, which, although it originated in China, has nevertheless become the dominant value system of and has had a tremendous impact on all East Asian countries.

V. QUANTITATIVE FUTILITY AND THE CONFUCIAN NOTION OF HEAVEN

(A) Tianming *(Heaven's Destiny)*

An illustrative example of a scientific and intellectualistic approach to ethical discourse is found in the statistical and quantitative methods employed to determine medical futility reviewed earlier. As we have discussed, these methods are not entirely value-free, have inherent limitations, and are at odds with the experiential approach to philosophy in the Confucian tradition which takes more seriously the patient's and the patient's family's traditions, values, emotional needs, and symbolic goals. Furthermore, Confucian philosophy would also be sceptical of the certainty and accuracy and hence the validity of applying statistical inferences derived from a group of patients to predict what may happen to a particular patient. There is an element in the Confucian conception of personhood which stresses the particularity of an individual, and this is due largely to the notion of *tianming* (Heaven's Destiny).

One of the most fundamental Confucian concepts of a person is the view that man is a member of the triad of Heaven, Earth, and Man. It is certainly true that Confucius himself avoided indulging in excessive speculation on the supernatural realm. He cautioned his disciples "to keep one's distance from the gods and spirits while showing them reverence ..." (*The Analects* VI:22; my translation) and questioned that "if man is so inadequate in his service to man, how does man expect to be able to serve the spirits?" (*The Analects* 11:12; my translation). On the other hand, one must not be misled to think that the transcendent realm is thereby completely excluded from Confucian thought. On the contrary, Confucianism does not doubt the existence of a transcendent Heaven (*tian*) which is very much involved in human affairs, and Confucian personhood includes a transcendental dimension which obligates human beings to comply to and not contradict Heaven's intentions and purposes. Admittedly, the notion of "Heaven" has gone through different stages of transformation from a predominantly theistic understanding to a naturalistic and later a moralistic notion. Confucius himself retained the theistic understanding of Heaven prevalent in the Shang (1751-1112 BC) and early Chou (1111-249 BC) dynasties, which held the strong view that human destiny is in the hands of Heaven. Confucius once said, "[t]he gentleman stands in awe of three things. He is in awe of the Decree of Heaven. He is in awe of great man. He is in awe of the words of the sages" (*The Analects* 16:8). Heaven is to be taken with utter seriousness, for, as Confucius also said, "when you have offended against Heaven, there is nowhere you can turn to in your prayers" (3:13). Mencius went so far as to say that "those who are obedient to Heaven are preserved; those who go against Heaven are annihilated" (*Mencius* 6A:7). In fact Heaven is believed to be such a strong supernatural force that it determines human Destiny and is beyond human control and manipulation. "It is Destiny if the Way prevails; it is equally Destiny if the Way falls into disuse. What can Kung-po Liao do in defiance of Destiny?" (*The Analects* 14:36). And when Confucius' most beloved disciple, Yen Yüen, died, he said, "Alas! Heaven has bereft me! Heaven has bereft me!" (11:9) indicating that Confucius understood Yen Yüen's death as nothing less than an act of Heaven. It is generally accepted in Confucian thought that one's longevity is decided by Heaven so that someone may enjoy a long lifespan while others may not. This idea of a specific Heaven's Destiny (*tianming*) for each person with one's lifespan determined for each particular individual is so deeply ingrained in Chinese popular thought that one should not expect a Chinese to accept easily a determination of medical futility based on certain statistical averages, which removes this element of particularity.

Yet, conceivably, if the determination of medical futility is interpreted in the context of the will of Heaven for that particular person, the notion of *tianming* may actually facilitate the acceptance of such a decision. This is particularly so if that decision is not statistically derived and is communicated sensitively and by an authoritative figure such as a physician (see below).

(B) Tiannen *(Heaven's Year)*

If a man's destiny is decreed by Heaven, it follows that one should seek to know and act in accordance to Heaven's will. In fact, in Confucianism, to understand Heaven's will is part of the process of the self-cultivation of a person. For this reason, Confucius said, "A man has no way of becoming a gentleman unless he understands Destiny" (*The Analects* 20:3). Mencius advanced a more moralistic interpretation of Heaven and identified it with the moral nature of man, which provided a more immanent way of knowing the will of Heaven. He said, "[f]or a man to give full realization to his heart is for him to understand his own nature, and a man who knows his own nature will know Heaven" (*Mencius* 7A:1). That is why for Mencius, there is an ethical approach to "know" and to "serve" Heaven. When a person fulfills his moral nature he is fulfilling the moral order of the universe, and in so doing, fulfills the command of Heaven; in fact, he becomes united with Heaven. Mencius went on to say, "[b]y retaining his heart and nurturing his nature he is serving Heaven" (*Mencius* 7A:1).

This Confucian idea of knowing and serving Heaven provides a much more active role in the interpretation of Heaven's destiny than the Taoist school of philosophy, which offers a purer form of natural determinism. That is to say, Confucianism allows a person to take a more active role to conform oneself to Heaven's destiny. Mencius said, "[t]hough nothing happens that is not due to Destiny, one accepts willingly only what is one's proper Destiny ... He who dies after having done his best in following the Way dies according to his proper Destiny" (7A:2). The suggestion of a "proper Destiny" evidently allows for the possibility of improper Destiny. Mencius went on to give two examples of improper Destiny: "It is never anyone's proper Destiny to die in fetters" and also "he who understands Destiny does not stand under a wall on the verge of collapse" (7A:2). The first example suggests that a person should cultivate one's character and the second example recommends that one should take appropriate actions to improve upon one's life situation in order to make sure that the Destiny that befalls one is proper, specifically in matters relating to death and dying. Furthermore, Mencius' teaching has been understood to mean that only if one has done one's best in following Heaven's Destiny can one consider one's death proper. In other words, if one's longevity is so determined by *tian* (Heaven), then it is man's proper responsibility to fulfill *tiannen* (Heaven's year), i.e. to fulfill the full measure of one's lifespan that Heaven intends one to have. Han Fei Tzu (d. 233 BC), a Confucianist with a more naturalistic inclination, said, "[c]omply to the years given by nature, one's life can be considered complete and fulfilled" (*Han Fei Tzu*, Bk.VI, my own translation). In light of this, a judgment on the medical futility of a procedure such as CPR (cardio-pulmonary resuscitation), which conceivably can stretch one's lifespan, may be considered a form of noncompliance to Heaven's proper destiny and should be viewed with suspicion. But if this teaching of *tian-nen* is taken into consideration, and a DNR (Do-Not-Resuscitate) order is justified on the basis that such a procedure in fact takes the patient beyond the patient's lifespan as destined by Heaven, then the notion of *tian-nen* may be used to improve the chance for the decision to be accepted.

VI. QUALITATIVE FUTILITY AND THE CONFUCIAN CONCEPT OF HOLISTIC PERSONHOOD

(A) The Confucian High View of the Body

The Western tendency to employ an abstract and intellectualistic approach to medical futility decisions is particularly evident in the qualitative definition of futility reviewed above. As Schneiderman *et al.* have argued, treatments that only preserve "continued biologic life without conscious autonomy" are deemed qualitatively futile (p. 950). In other words, when a person's rational faculty is destroyed or sufficiently diminished, the treatment that the person receives is no longer considered beneficial. But this conclusion is only deemed reasonable if one also subscribes to the anthropology of Schneiderman *et al.* in which the physical body is evidently only a negligible part in "the patient as a whole." Such a low view of the body is at odds with the Confucian understanding of the person. To start with, a very important reason for the Confucianists' high regard for the physical body is due to their predominantly "this-worldly" orientation (which we will discuss in some detail later in the section on immortality and death). Suffice it here to say that Confucian thought is a "here-and-now" philosophy primarily concerned with how to live an earthly life through which Chinese people are made aware of the importance of the physical body. As one commentator observes, even after the introduction to China of the Buddhist ideas of "afterlife" and "rebirth", the Chinese people are not particularly interested in becoming Buddhas who are "free from sin, sorrow and suffering ... [and] escape transmigration ... This appeals to some very devout Buddhists, but not to the Chinese people in general. They want to enjoy the present life or to accumulate merit so as to enjoy a happy and fortunate existence after rebirth. The Chinese ... prefer life in this world" (Graham, quoted in Watson and Rawski, p. 193). And this means bodily life.

The Confucian high view of the physical body is ultimately a reflection of the holistic conception of human personhood held by this tradition. There are two levels to this holistic conception. In the first place, from the perspective of the Heaven-Earth-Man triad, Confucianism holds that humans and the cosmos constitute one holistic entity and recognizes that a person's existence on earth and his connectedness with nature is mediated through the physical body. In fact, human beings, especially in their physical aspects, share so much in common with nature that the human body and the natural physical environment are believed to be inseparably and interdependently related.[2] Through the medium of the vital material force (*qi*),[3] the human person, including even the various organs of the human body, is connected to nature and the cosmos (Huang Ti, pp. 105). This explains why Chinese medicine puts so much emphasis on the importance of the environment for one's physical health. The well-being of the physical body reflects a state of desired harmony between the person and nature, thereby affirming the unity of person and the cosmos.

But the most important reason why the body is accorded with a high place in Confucian thought is due to a second level of understanding of holistic personhood. Confucian anthropology holds that a human being is constituted by three components: *xin*[a] (heart, will or mind), *qi* (vital force), and *ti*[a] (body).[4] The *xin*[a] possesses intentionality, the *ti*[a] is the visible and locomotive aspect, while the *qi* mediates and regulates the two. Confucian anthropology holds a holistic understanding of the human person and does not consider any one of these three aspects to have priority over the other two. In fact it is held that all three components are given to human beings by Heaven and are dynamically related to and interdependent on each other to form a holistic organic entity which is the human being. Given this view, to determine futility and withhold treatment from a person just because rational choices and autonomy are no longer possible and all that is left is the mere physical body of that person will not be viewed with favor from a Confucian perspective.

This Confucian high view of the physical body creates something of a paradox, especially because the Confucian conception of an ideal person is a moral sage. For Mencius, a sage is a person whose moral character has been perfected, and since no one is born a sage, to become a sage involves a process that includes all three aspects of the person. According to Mencius, before sagehood is attained, a person is incomplete, striving to be perfected and fulfilled. The person is a being always in the process of becoming. This implies that all the three aspects of the person: the heart/mind (*xin*[a]), the vital force (*qi*), and the physical body (*ti*[a]) are imperfect, waiting to be perfected. This can be seen in three passages from Mencius. For the heart, Mencius said, "[f]or a man to give full realization to his heart is for him to understand his own nature, and a man who knows his own nature will know Heaven" (*Mencius* 7A:1). In other words, self-knowledge is the means to progress toward perfection of one's mind, and only a few are able to attain perfect self-knowledge in their life-time. As for *qi*, the vital force, Mencius said, "if the will is concentrated, the vital force [will follow] and become active. If the vital force is concentrated, the will [will follow it] and become active ... And I am skillful in nourishing my strong, moving power [force, *qi*]" (2A:2). To nourish one's *qi* is one of the most important lessons of self-cultivation in Confucian thinking. And finally for the body, Mencius said, "[o]ur body and complexion are given to us by Heaven. Only a sage can give his body complete fulfillment" (7A:38). This passage is interpreted to mean that it takes the moral character of a sage to actualize and fulfill the body; in turn, Mencius taught that the qualities of sagehood are exhibited through the body. "That which a gentleman follows as his nature, that is to say, benevolence, rightness, the rites (propriety) and wisdom, is rooted in his heart, and manifests itself in his face, giving it a sleek appearance. It also shows in his back and extends to his limbs, rendering their message intelligible words" (7A:21). So, the whole body manifests the glory of the moral person.

This high regard for the physical body is the main reason why organ donation and post-mortem examination are so unpopular since both require tampering with the body.[5] Also, most Chinese knowledgeable in Confucianism believe that, when a person has lost his mental faculty (*xin*[a]) he may have lost a very important component of the person, but the rest of the person remains. The person is not thereby reduced to, in Schneiderman *et al.*'s terms, "some portion of the patient's anatomy, physiology or chemistry ..." (p. 950) for which all medical treatments would predictably yield "qualitatively poor results." This is not to suggest that a Confucianist would insist to maintain a (total) brain dead human body with advanced life support, only to point out that the determination of futility based on psychological criteria must be negotiated sensitively with the Confucian conception of the body. This also suggests that a Confucianist would encounter considerable difficulty to accept futility determinations for one who is in a so-called "permanent vegetative state." On the other hand, this high regard for the physical body held by Confucianists may also work in favor of not consenting to futile treatments especially when these involve extensive surgical interventions and subject the body to disfigurement, dismemberment, or other forms of mutilation.

(B) The Value of Physical Suffering

For Mencius, the possibility of improving on the three aspects of the person is provided by the original moral goodness inherent in all human nature (See *Mencius*, 6A:6). When the good nature is discovered, cultivated, and nourished, complete moral personhood (sagehood) is attained. The body, instead of being dismissed or rejected as being of lesser value, is in fact elevated because in the Confucian conception of sagehood, the perfected moral personhood is both derived from and exhibited through the body. This is made clear by a particularly influential passage in which Mencius said, "[t]hat is why heaven, when it is about to place a great burden on a man, always first tests his resolution, exhausts his frame and makes him suffer starvation and hardship, frustrates his efforts so as to shake him from his mental lassitude, toughen his nature and make good his deficiencies" (6B:15). This teaching has had tremendous impact on the Chinese mind with regard to the meaning of physical suffering. For this reason, a common belief among Chinese people is that "It is more preferable to hang on to life, even a "bad life," than to give oneself up to a good death." For to give oneself up to a good death is an attempt to escape the "character cultivation" that comes with suffering which is a necessary step towards sagehood. So even when life is nearly unbearable with much suffering, a Confucianist may see in it a positive, purgative, and pedagogic value and is motivated to endure it.

Hence, Confucianism does not share the prevalent Western viewpoint that when a person's survival requires extensive bodily care with no other life goals, then medical treatments are considered futile and no longer beneficial (Schneiderman, *et al.*, p. 953). Such quality of life criteria betrays a clear bias against the physical body for which Confucianism retains a much higher view, and dismisses any value in physical suffering which Confucianism regards as pedagogically meaningful.

(C) The Confucian Concept of Death and Immortality

One of the best examples of Confucianism's "this-worldly" orientation is its very scarce teaching on death. Confucius himself, though not an agnostic, showed very little interest in the event of death. He was much more concerned with the moral and sociopolitical structures of this world than with postmortem realities of deities and spirits. Once pressed by a student's question concerning death, Confucius evasively answered, "if we do not yet know about life, how can we know about death?" (*The Analects* 11:11). This is in contrast to the West where despite the increasing secularization, one may still be able to say that for most people the rationalism of the Enlightenment has not eliminated the idea of some form of postmortem survival of the soul, an idea handed down from Plato and continued through the Christian era. Although the nature of the immortal soul may have been different in the hands of different philosophers, Western culture continues to retain an eschatological perspective in life, and the majority of the Western people continue to look towards a postmortem existence, however vague that may be, and derive a sense of peace and hope when confronted with the inevitability of death (See Choron). This kind of teaching is notably absent in the Confucian tradition.

Because Confucian teaching is primarily concerned with the affairs of this world, the only notion of immortality in Confucian thought is a form of metaphorical or "historical" immortality which does not involve an actual postmortem existence for the person. Confucianism teaches the "doctrine of the three kinds of immortality," which has had the most important impact on the Chinese attitude towards death. This doctrine is recorded in the ancient history book of *Tso Chuen* (compiled around the 5th century BC) which states clearly that only three things are immortal in the world: virtues, successes, and writings.

> When Muh-shuh (P'aou) went to Tsin, Fan Sëuen-tsze met him, and asked the meaning of the saying of the ancients, "They died but suffered no decay," ... Muh-shuh said, " ... I have heard that the highest meaning of it is when there is established [an example of] virtue; the second, when there is established [an example of] successful service; and the third, when there is [an example of wise] speech. When these examples are not forgotten with length of time, this is what is meant by the saying 'they do not decay'" (*Tso Chuen: Duke Seang*, p. 507).

For Confucianists, the highest form of virtue is *ren*, and in comparison to achieving *ren*, which belongs to the immortal category, life pales in significance. For this reason, Confucius said, "for gentleman of purpose and men of benevolence [*ren*] while it is inconceivable that they should seek to stay alive at the expense of benevolence, it may happen that they have to accept death in order to have benevolence accomplished" (*The Analects* 15:9). In other words, if *ren* is achieved, human personhood is fulfilled and death can be accepted. But this also means that when *ren* has not been achieved, there may be a reluctance to accept death even when death is immanent and inevitable. To leave this world without attaining a certain degree of virtue is to have lived in vain. Only the attainment of the Way (*dao*) of life which is embodied by the virtue of *ren* releases the person from this duty. For this reason, Confucius said, "In the morning, hear the Way; in the evening, die content!" (IV:8, Chan, p. 26). For similar reasons, accomplishments in civil services or academic writings are important determinants of one's acceptance of death. "The gentleman detests not leaving a name behind when he is gone" (*The Analects* 15:20), and to die an "ordinary man ... is something worth worrying about" (*Mencius* 4B:28). In light of this, when confronted with decisions of futility determination, one may better understand why a Confucianist is likely to be influenced by whether or not he/she has accomplished one, two, or all three of these immortal goals. For a Confucianist, since death terminates all possibilities of further human activities, death is primarily evaluated in terms of these accomplishments. It is a "worthy death" or "proper destiny" only when most, if not all, of one's duties of life have been properly fulfilled. Any resistance to acknowledge medical futility may mean some unfinished business perceived by the patient, and may explain some desperate and seemingly irrational attempts to seek medical treatment to recover from the illness or to extend life to the last breath in order to complete one's life tasks.

On the basis of this analysis of the Confucian concept of holistic personhood, one may have gained the impression that there exists in the Confucian tradition a proclivity to reject futility determinations, particularly when they are assessed on the basis of psychological/qualitative criteria. This impression is correct only when such determinations are made with reference to the person exclusively as an individual self or as an independent member of the Heaven-Earth-Man triad living in isolation in the society. But this is an oversimplified view of Confucian personhood because a person as a "being-in-oneself" is an alien notion in Confucianism, which has a concept of personhood that always sees a person as a "being-in-relations." In the context of this Confucian notion of social personhood, futility considerations may take a different turn. As we will see in general the Confucian concept of social personhood exists in tension with the Confucian concept of holistic personhood, and in general, the former will outweigh the considerations arising from the latter.

VII. CONFUCIAN SOCIAL PERSONHOOD AND DECISION MAKING IN MEDICAL FUTILITY

Since the modern western conception of personhood has emphasized the ability to exercise autonomous choices as *sine qua non* for qualification as a person, the principle of respect for autonomy has also become one of the most important principles in biomedical ethical deliberation. As we have seen, in the various deliberations with regard to whose opinion should be decisive in the judgment of medical futility, the issue is often divided by two opposing "poles" of autonomy, that of the patients and that of the providers. The confusion, or at least the inefficiency, that is associated with the application of this principle is quite remarkable. One only needs to be reminded of two pairs of cases. The first pair refers to the Karen Ann Quinlan case in 1976 and the Nancy Cruzan case in 1990 in which families petitioned on the basis of respect of autonomy to withhold life-sustaining treatments that health care providers did not consider to be futile. The second pair refers to the "Baby L" and Helga Wanglie cases in which families failed in their bid to continue intensive treatments which the hospitals deemed them futile. It is clear in these two pairs of cases that the principle of patient autonomy is interpreted quite differently, with it being upheld in the first pair and overridden in the second pair with no convincing reason as to why it would be either. In this section of the paper, four key Confucian concepts pertaining to "person-as-being-in-relations" will be discussed which we believe shape these decision making issues quite differently: (i) social personhood; (ii) the principle of *ren*; (iii) social hierarchy the doctrine of "*zhong-yong*"; and (iv) filial piety. These concepts will be presented first, and their application to futility issues will be discussed together at the end.

(A) Social Personhood in Confucianism

If autonomy is the centerpiece of the western conceptualization of personhood, it may be said that the central characterization of Confucian personhood is man's social and relational nature. The early Confucian thinker Hsün Tzu (298-238, BC) recognized that human beings are distinguishable from animals because of man's social capacity. He said, "[m]en are not as strong as oxes, nor do they run as fast as horses, yet how is it that oxes and horses are being mastered by men? That is because men are capable of social organization and animals are not" (*Hsün Tze* 7:9, my translation). In the Confucian tradition, a person is never seen as an isolated individual but is always conceived of as a center of relations. Even though it may seem to many that the bulk of Confucian ethical teaching pertains to the self-cultivation of an individual person, it should be remembered that the whole process is carried out in a social context and for the purpose of fulfilling social responsibility. Even when the cultivation of subjectivity as a holistic person has been emphasized, especially by later Neoconfucianists of the Sung Dynasty (960-1279 AD), this so-called subjectivity is located within a social collective subjectivity. Any subjective consciousness that a person has been made aware of through self-cultivation is more a reflection of a collective subjectivity of the community than it is of his own subjectivity. The purpose of self-cultivation is never to establish the independence or individuation of the subject, but to promote and maintain the collective harmony of the community. It may be asserted that the entire Confucian program for self-cultivation is to emphasize the social nature of man. To actualize man's nature is to fulfill man's human relatedness. In Confucianism, a person is always a "person-in-relations." Mencius said, "Slight is the difference between man and the brutes. The common man loses this distinguishing feature, while the gentleman retains it. Shun understood the way of things and had a keen insight into human relationships. He followed the path of morality" (IVB:19, p. 165).[6] So the ability to have human relationships is what differentiates human persons from animals, and what is crucial for a human person is to be self-consciously aware of "a keen insight" in human relationships and so "by nature" follow "the path of morality." Mencius was cognizant of the danger that human beings may lose this consciousness of human sociality and relationships and thereby degenerate into behaviors not significantly different from animals. He said, "the multitude can be said never to understand what they practise, to notice what they repeatedly do, or to be aware of the path they follow all their lives" (*Mencius* 7A:5). This is why Mencius repeatedly emphasized the importance of education and moral cultivation. When a person is in possession of "a keen insight into human relationships," he becomes a fulfilled and actualized true person a sage. Mencius said, "[t]he [carpenter's] compass and square produce perfect circles and squares. The sages exhibit perfect human relations" (IVA:2, my translation). It may be said that these relations are of the ontological category in Confucian anthropology. They both constitute and complete personhood.

(B) The Principle of Ren

The doctrinal core of Confucianism is the principle of *ren*, which has been variously translated as benevolence, humanity, humanness, compassion etc. In *The Analects* alone, the term "*ren*" appears 105 times, more than any other term including heaven, ritual, and filial piety. All the other Confucian ideals: humanity, personality, morality, and political ideology are deduced from this central doctrine. Depending on the different contexts in which the term is being used, it can be a challenge to try to understand its exact meaning. To start with, *ren* is a relational term and is used to describe the relational/social nature of man. The ideograph for *ren* is composed of two characters "man" and "two", denoting that whatever else the term may mean, its meaning is intended to be accomplished through human relationships. As such, *ren* is not understood to be merely one of the many virtues, but is an all-inclusive term which encompasses all virtues that guide conduct in human relationships. When one of Confucius' disciples asked him about *ren*, the Master said, "There are five things and whoever is capable of putting them into practice in the Empire is certainly *ren* ... They are respectfulness, tolerance, trustworthiness in word, quickness and generosity" (*The Analects* 17:6). In this sense, *ren* embodies the totality of a virtuous Confucian person. Mencius said, "*ren* means 'man'" (*Mencius* 7B:16). In other words, Confucianism understands *ren* as a universal virtue which is intrinsic to human nature. Mencius referred to human nature as the heart of man, and he identified *ren* with one's heart: "benevolence (*ren*) is the heart of man ..." (6A:11) and "a gentleman retains his heart by means of benevolence ..." (4B:28). When *ren*, the human nature, is fulfilled in a sage, it is found in everything he does and at all times. Confucius said, "the gentleman never deserts benevolence (*ren*), not even for as long as it takes to eat a meal. If he hurries and stumbles one may be sure that it is in benevolence that he does so" (*The Analects* 4:5). Furthermore as part of human nature, *ren* is more than a set of moral rules and precepts. Rather, it is out of *ren* that moral conducts are derived. For this reason, Mencius said, "Shun ... followed the path of morality. He did not just put morality into practice" (4B:19). In this sense, *ren* is prior to precepts. Confucius once said, "when faced with the opportunity to practise benevolence (*ren*), do not yield precedence even to your teacher" (*The Analects* 15:35). But as we will see later, the rule of teachers having precedence over pupils is otherwise quite rigidly enforced in Confucian ethics.

Yet the principle of *ren* is also understood and used as an ethical norm that serves the most basic function of regulating human relationships in Confucian thought. Confucius taught that all human relationships should be guided by *ren*. When his pupil Yen Yüen asked him about *ren*, he said, "to return to the observance of the rites through overcoming the self constitutes benevolence (*ren*)" (*The Analects* 12:1). Here the rites may simply be understood as a set of protocols for human relations. Confucianism recognizes that as an ethical norm, the content of *ren* is love. Once asked by his pupil Fan Ch'ih about *ren*, Confucius said, "Love your fellow men" (12:22). In Mencius' writing, the theme of love is also found to be central to his understanding of *ren*: "the benevolent man loves others ..." (*Mencius* 4B:28); "a benevolent man loves everyone ..." (7A:46); and he "extends his love from those he loves to those he does not love" (7B:1). In concrete moral terms, *ren* as love is expressed negatively as "do not impose on others what you yourself do not desire" (*The Analects* 12:2); and "what you do not wish others to do to you, do not do to them."[7] The end result in view is the elimination of evil, for as Confucius said, "if a man were to set his heart on benevolence, he would be free from evil" (4:4). Positively, *ren* as love is to be understood in this way: "A man of humanity (*ren*), wishing to establish his own character, also establishes the character of others, and wishing to be prominent himself, also so helps others to be prominent" (6:28). Confucius elaborates this in terms of how one's practice of *ren* may benefit others. "The superior man is one who cultivates himself with seriousness ... so as to give all people security and peace ..." (14:45). Mencius envisioned that through the extension of love, the whole community will live in peace. He explained this to King Hsuan: "Treat the aged of your own family in a manner befitting their venerable age and extend this treatment to the aged of other families; treat your own young in a manner befitting their tender age and extend this to the young of other families, and you can roll the Empire on your palm ... . Hence, one who extends his bounty can tend those within the Four Seas; one who does not cannot tend even his own family" (*Mencius*, 1A:7). The end result in view is the establishment of a peaceful Kingdom of unity (*Datong*[8] [Grand Union]; see *The Li Ki* [The Book of Rites], Book VII, Section 1, Müller, pp. 364-366).

(C) Social Hierarchy

In the actual implementation of the principle of *ren* in human relations in a society, Confucianists have adopted a rigidly hierarchical structure which has been known as the "Five Relations." It was Mencius who first proposed this structure when he said that "this is the way of the common people; once they have a full belly and warm clothes on their back they degenerate to the level of animals if they are allowed to lead idle lives, without education and discipline. This gave the sage king further cause for concern, and so he appointed Hsieh as the Minister of Education whose duty was to teach the people human relationships: Love between father and son, duty between ruler and subject, distinction between husband and wife, precedence of the old over the young, and faith between friends" (*Mencius* 3A:3). The principle of *ren* plays its part differently in each of the five relationships to provide the different affections or virtues that should bond the relations. For parents and those in superior positions, the predominant affection is kindness and rightness. For children in relation to their parents, "*xiao*" (filial piety) is emphasized as the appropriate attitude, while "*ti*[b]" (brotherly respect) is the proper disposition one should have towards one's older siblings. The emphasis on hierarchy and the distinctions among members within the family is illustrated by the fact that in the *Er Ya*, the oldest dictionary of the Chinese language compiled around the 2nd century BC, there are more than one hundred terms for various family relationships. "*Ming-fe*n," which literally means "name-distinction," is a very important idea in Confucian thought that stresses and dictates a person's specific position with respect to other people in the family and is accompanied by the specific privileges and obligations associated with that position. This stems from the Confucian idea of "Rectification of Names" (*zheng ming*).[9] Confucian teaching emphasizes that when everybody accepts one's role and discharges one's duty in accordance with the various "named" roles within the family, family order will be secured. As the extension of the differentiated and hierarchically arranged roles within the family, a similar hierarchy can be justified in a wider social context in which "*zhong*" (faithfulness) is practised between an emperor and his subjects, and "*shu*" (altruism) is practised between a superior and his inferiors and between a senior and his juniors. Together, *xiao* (filial piety), *ti*[b] (brotherly respect), *zhong* (faithfulness), and *shu* (altruism) provide the concrete expressions of *ren* encoded in the *li*[b] (rites), and form the basis of social solidarity and harmony. Confucianism emphasizes the need to live harmoniously in the society. Confucius' pupil Yu Tzu is quoted to have said, "Of the things brought about by the rites, harmony is the most valuable. Of the ways of the Former Kings, this is the most beautiful" (*The Analects* 1:12).

Within the context of a clearly differentiated and hierarchically arranged social structure as the basis of social harmony, Confucianists have developed the doctrine of "*zhongyong*" to assist in achieving harmony in a Confucian society. "*Zhong*" literally denotes moderation and appropriateness, while "*yong*" means universal and harmonious. Together the term "*zhongyong*" implies that "there is harmony in human nature and that this harmony underlies our moral being and prevails throughout the universe" (Chan, p. 96). To ensure a smooth resolution of the differences between people, the practice of "*zhongyong*" is considered by Confucius to be a virtue of a sage. A passage in the Doctrine of the Mean of the Book of Rites says, "Equilibrium is the great foundation of the world, and harmony its universal path. When equilibrium and harmony are realized to the highest degree, heaven and earth will attain their proper order and all things will flourish" (Chan, p. 98).

(D) The Confucian Family and the Principle of Xiao

From the perspective of establishing sociopolitical stability and harmony in a society, Confucianism has always maintained that the family is the most fundamental unit in social organization, with the father-son couplet being the key relationship. The strict conduct of a son towards his father, as "*xiao*" (filial piety), is easy to establish since the bond is reinforced by "blood ties." Once this relationship is established, the authoritative status of the father can be readily and legitimately extended to other elders in the extended family, then to the local community with thinner "blood ties," and ultimately to the rest of the society even when there are no "blood-ties." To justify this hierarchy and the proposed appropriate affections, Confucianism teaches that these affections are nothing less than expressions of one's intrinsic human nature as *ren*. Confucius said, "Filial piety and brotherly respect are the root of humanity (*ren*)" (*The Analects* 1:2). Mencius wholeheartedly agreed in saying that "the content of benevolence (*ren*) is the serving of one's parent" (*Mencius* 4A:27). Since filial piety is elevated to become a matter of *ren*, when Meng Yi Tzu asked about filial piety, Confucius simply answered, "Never fail to comply" (*The Analects* 2:5). More specifically, Confucius stated that "When your parents are alive, comply with the rites in serving them; when they die, comply with the rites in burying them and in offering sacrifices to them" (2:5).

One of the most concrete expressions of the Confucian teaching of filial piety is in terms of the physical well-being of both the parents and the children. Children from a Confucian family are supposed to be very concerned about their parents' health. When a pupil of Confucius asked about filial piety, Confucius' answer was that he should "especially be anxious lest parents should be sick" (2:6). Mencius gave the example of Shun as an example of a good son because "At the age of fifty, he still yearned for his parents" (6B:3). Children should always be attendant on their parents' well-being, for as Confucius taught: "while your parents are alive, you should not travel too far. If you do travel, your whereabouts should always be known" (IV:19, p.33).

Confucianism also mandates that one should treat one's own physical well-being as an expression of filial piety. One's body is literally understood to be one's parents' body and hence must be treated with the utmost respect. A passage in the Book of Rites (*Li Ji*) states that "The body is that which has been transmitted to us by our parents [literally translated as our parents' body left to us], dare any one allow himself to be irreverent in the employment of their legacy?" (p. 226). For this reason, Chinese people consider maintaining one's health as a duty of filial piety. As the Book of Filial Piety (*Xiao Jing*) clearly states, "Our bodies to every hair and bit of skin are received by us from our parents, and we must not presume to injure or wound them: this is the beginning of filial piety" (Ch. 1, p. 466). This concern not to harm one's body so as to avoid violating the rules of filial piety can become quite excessive. There is a story in *The Analects* about one pupil of Confucius, Tseng Tzu, who became seriously ill and summoned his disciples to examine his hands to be sure that his body, which he regarded as that of his parents, was not injured or mutilated due to his own illness (Confucius, VIII:3, p. 69). One can see from these teachings why sometimes it is so difficult for health care providers to persuade those from a Confucian background to agree to terminate medical treatments and allow the illness of the patient (either oneself or one's parents) to take its own course.

(E) Implications for Decision Making

These four Confucian concepts of social personhood, *ren*, social hierarchy and filial piety provide several different perspectives on the issues of medical futility. We will consider their impact particularly on the decision making processes of both the patient(s) and the physician(s).

From the patient's point of view, a strong sense of individualism and an imperative to respect an individual's right to self-determination simply are not present in the Confucian social organization. In turn, the assumption that the patient should be told of the diagnosis and prognosis and should be the one to make medical decisions is challenged by the Confucian concept of social personhood. When a Confucian Chinese becomes sick, his behavior and expectations are closely tied to a strong sense of personal identity with and dependence on his family. For example, it is not shameful for a Chinese elderly sick person to be dependent on his children, it is rather a privilege to which he is entitled. According to *The Book of Rites*, the rules are: "When a ruler is ill, and has to drink medicine, the minister first tastes it. The same is the rule for a son and an ailing parent" (Bk. 1, Part III, Sec. 2, p. 114). In conformity to the accepted arrangements within the Confucian social hierarchy, the elderly sick person can expect to be cared for, and his role includes the privilege to be relieved of a large share of personal responsibility, including most of the decision making processes of his own medical care, even though he may be rational and competent. Family members, especially the patient's children, are expected to take over that responsibility and assume the various roles of being children, caregivers, protectors, and surrogate decision-makers. While in the West family input is usually not decisive, in Confucian culture, this input is often determinative, making the conventional requirement of patient informed consent, as prevalent in the West, not easily enforceable.

If Confucian culture generally excuses elderly patients from being informed of their medical condition and sharing in the responsibility of their own medical care, it specifically shields patients from being apprised of a terminal illness, and/or the judgment that all subsequent treatments are considered futile. For the prevalent Western bioethical practice, it is both appropriate and necessary to provide the patient with all the medical facts, including the fateful ones, directly, fully, and truthfully. However, for a Confucianist governed by the rule of filial piety it may be considered to be morally inexcusable to disclose the news of a terminal illness to an elderly patient as it may add further harm to the patient. To fulfill filial piety, dying elderly patients must be protected from fateful news.

The central role the family plays in a Confucian society also exerts important influence in a number of different ways directly on medical futility decisions. In general, because the Confucian tradition has such a high regard for the elders in its social organization, one can expect the family to try most of everything possible to save or extend the life of a family elder. However, under severe financial constraints, when the cost of continuing medical treatments is so high that the livelihood of the rest of the family members is being jeopardized, especially when the likelihood of medical benefit is slim, it is not uncommon for elderly patients to request the termination of further treatments. The principle of *ren* requires family elders to show compassion to the younger generations in the family. To drain family members physically and emotionally because of a lengthy illness from which a family elder is not likely to recover can hardly be considered to be an act of compassion. To squander a large sum of money or to incur heavy family debts in pursuit of futile treatments also would not be considered right or appropriate. The idea of doing what is "appropriate and right" is called "*yi*[a]" in the Confucian tradition, and is an important ethical expression of *ren*. To fulfill *ren* and *yi*[a] in the context of a Confucian family under stress, an ailing elderly person is likely to concur with a physician's opinion of medical futility.

Sometimes, the principle of *ren* expressed through the rule of filial piety plays its hand in a different way. When confronted with inevitable death, an elderly person may be willing to accept the determination of medical futility, but his children are found to be reluctant to grant him this wish because they do not want to be deprived of the remaining opportunity to show their filial piety. It is a strong Confucian idea that filial piety is best expressed while a parent is alive and so it is a human tragedy when a son wants to show his filial piety but his parent is not alive to receive it. So to extend the life of an ailing patient is to extend the opportunity to show filial piety. This alone is a common reason for the children of a sick parent to reject the determination of medical futility and to request everything be done for their dying parent, sometimes even when great suffering result from continued medical intervention. Regrettably, much of these practices that are based on filial piety are self-serving, benefiting the children of the patient rather than the patient himself. At best, it can be seen as an expression of *ren* in order to establish the virtuous character of the self, and at its worst, it is merely an attempt to avoid a strong sense of guilt and shame. In the Confucian social hierarchy, shame is a particularly strong emotion. To fail to discharge one's assigned role and duty within the "five relations" is to be guilty of upsetting the hierarchical structure, and the punishment is to be treated with shame. Hence, the fear of shame due to the failure to show sufficient filial piety to a parent or family elder can be a strong reason to request life-extending measures. To agree to stop life support may be considered a shameful "face-losing" decision and must be avoided at all cost. But since shame is the product of a perceived failure in abiding by the rule of filial piety within a rigid social hierarchy, conceivably, this problem of equating filial impiety and shame with the acceptance of medical futility can be circumvented if an authoritative figure within the social hierarchy speak out to correct such a misperception. Often, this figure is found in the person of the physician involved.

From the physician's point of view, both his role and function are subject to the regulatory principles and rules associated with the Confucian concepts of social personhood and its associated societal hierarchy. To start with, Confucianists believe that the principle of *ren* should be the regulatory principle for the physician's character as well as his conduct just as it is for other members of the community. But because benevolence is particularly important for the profession of medical care, Confucianism specifically labels the profession of medicine as "the profession of *ren*" (See Ma, pp. 487-490) and the first duty of this "profession of *ren*" is not to inflict harm. Mencius said, "there is no harm in this. It is the way of a benevolent man" (1A:7). This closely parallel to the Western Hippocratic tradition of *primum non nocere* and affirms that the Confucian principle of *ren* favors the contemporary biomedical principle of beneficence (See Beauchamp and Childress) as the predominant guiding principle in biomedical ethics. Particularly, because Mencius taught that a compassionate heart is part of human nature (2A:6) and one fulfills the virtue *ren* by actualizing the virtuous human nature, there is a strong motivation for physicians to be compassionate in the conduct of their medical profession. For this reason, the highest compliment one can give to a physician is to say that he "has attained a *ren* heart and practised *ren* in his profession." And because *ren* is also the basis for all the other moral norms, a Confucian physician will usually follow the inclinations of the family to satisfy all the other social protocols, particularly the requirements of filial piety as perceived by the patient's family. On the other hand, according to the hierarchical organization in a Confucian society, the physician is in a social position to override the decision of a patient and/or his family if he so inclines. Even though the exact position of the physician is not specified in the Confucian scheme of the "five relations," traditionally physicians command respect equivalent to the parents and elders. There is a common saying in Chinese, "physicians have the heart of parents," indicating that people expect physicians to act as parents do, benevolently and compassionately. Physicians, therefore, are accorded a level of respect commensurate with elders and parents. Another interesting observation is that in the Confucian tradition, the status of the physician is second only to the Emperor's minister. In fact, from the Sung Dynasty onwards (960-1279 AD), the first vocational choice for a Confucian scholar was to become a minister, followed by the aspiration to become a physician (See Ma, pp. 476-477). With this elevated status, it is not uncommon for Chinese physicians to adopt a paternalistic attitude in their interactions with patients. They will not find it ethically troubling to withhold information from patients if they feel that it is not beneficial for them. In this regard, this paternalistic tendency acts synergistically with the inclination of the patient's family not to disclose ill-fated information.

The superior position of physicians in the Confucian hierarchy of relations also tends to give physicians priority in decision making, particularly those which are considered to be primarily medical in nature. While in the West, physician priority has sometimes been advocated by appealing to professional expertise (Murphy, p. 2099), moral integrity (Tomlinson and Brody, p. 1278), and autonomy (See Schneiderman, *et al.*, p. 953; Truog, *et al.*, p. 1562) , but in the Confucian tradition physician priority can be assumed because all three qualities are embedded in the hierarchically-arranged social organization. It is much easier for physicians to prevail over their patients from this authoritative position. Moreover, in keeping with the Confucian tradition, patients and their families find it quite natural to yield to the opinion of their physician.

In a homogeneous Confucian society, because patients and physicians basically subscribe to a similar set of values and beliefs, conflicts between patients and providers are relatively rare. Should the uncommon event of a major difference in opinion arise, instead of confrontation and antagonism, both parties usually show a strong desire to maintain harmony as it is constitutive of the Confucian teaching. In *The Analects* it is said, "Of the things brought about by rites, harmony is the most valuable" (1:12). By following the principle of "*zhong-yong*", both physicians and patients alike will be willing to avoid extreme attitudes and actions and adopt an accommodating posture towards each other in order to reach a compromise or settlement. Arbitration outside of the patient/family-physician relationship is seldom necessary, if ever, even for an issue as sensitive as medical futility.

VIII. CONCLUSION

The literature on medical futility that we have reviewed in this paper shows that to determine whether or not certain treatments are futile is far from being a value-neutral, objective, technical procedure. Instead it is closely intertwined with other human values having to do with our culturally bound concepts and beliefs relating to health, illness and death. Above all, medical futility issues expose our respective substantive visions of what it means to be a person. The liberal humanist view of the modern West brings to the "futility" debate a specific vision of the person as an autonomous chooser with full control over his own history and destiny. As in Nancy Cruzan's case, such autonomy may even be asserted on one's behalf by others when one is no longer able to assert one's own wishes (Crigger, 1990), for the purpose of safeguarding the patient against overtreatment by over-zealous, paternalistic physicians. But it has become increasingly obvious that such an account of personhood is inadequate to resolve all the issues surrounding futility determination. The Helga Wanglie case serves to illustrate just that: when in the name of patient autonomy Helga's family demands "non-beneficial" treatment, the community (hospitals and caregivers) feels obligated to oppose the patient's "autonomy" and to deny her care that is deemed "futile" (Miles, 1991). When there is no substantive agreement on what it means to be a person, such contradictions are to be expected and the process of shared decision making may at best offer a kind of procedural relief in dilemmatic situations. Until there is a *re-visioning* of the predominantly psychical (rational, autonomous) understanding of personhood in the West, most "futility" discussions can be expected to be themselves largely futile, ending up in a collision of several individual autonomous wills.

On the other hand, the Confucian understanding of personhood is radically different from that of the modern West. Instead of seeing a person as an independent, individualistic, and autonomous self, a Chinese person sees himself or herself as a member of the Heaven-Earth-Man triad, inseparably connected to and mutually dependent on the transcendent heaven, the natural environment, and above all the community in which the person lives and from which the person derives his or her identity. In confronting a death and dying situation, a Confucian Chinese draws on this rich and multi-dimensional notion of a person to make "futility" decisions. The Confucian account of personhood, which recognizes the transcendental dimension of the will of Heaven, may incline a person to resign to a more docile disposition of accepting the natural course of events. Its understanding of the personal significance of one's physical body reminds us that many of the "futility" determinations based on quality of life criteria are actually expressive of a Cartesian formulation of personhood. This formulation dichotomizes the person into a mechanical body and an intellectual soul, and it does not, according to a Confucianist perspective, take human embodiment seriously enough. Lastly, the Confucian concept of relational personhood, based on the principles of benevolence (*ren*) and appropriateness (*yi*[a]), makes social stability and harmony in human relations paramount in human communities. The teaching of *"zhong-yong"* specifically calls for a "give-and-take" posture and is against taking extreme positions in any human transaction. In the context of the "futility" debate in the West in which issues of economics and resource allocation inescapably enter into the picture, a relational understanding of personhood emphasizes that the interests of an individual and the community of which the same individual is a member ultimately can not be separated and must be carefully and sensitively balanced. In sum, the Confucian understanding of personhood has a great deal to contribute towards the West's *re-visioning* of its notion of personhood.

In the context of a cross-cultural clinical setting, this discussion has put the cultural differences between the eastern and western approaches to the medical care of terminal patients in sharp focus. Unfortunately, this cultural difference is not always well recognized, and the western medical perspective is sometimes assumed to be culturally universal, common to all people. But, as one observer points out, this presumed ethical agreement is more apparent than real, and "is a result of the diffusion of western ethics with western medical technology and medical social organization" (Kunstadter, p. 290), rather than ethical homogeneity. This unfortunate scenario often leads to painful confrontations when health care providers, in the process of decision making and health care delivery, fail to meet the minds of non-western health care recipients who are influenced by beliefs and values significantly different from those that inform western ethics. This is especially true for the younger generation of doctors, who think in the context of western medicine and erroneously believe that western values are the only values guiding ethics. We believe that patients from non-western cultures are entitled to have their socio-cultural background considered and their specific needs responded to. It is illogical and unsafe to treat all people with different cultural backgrounds alike. In a multi-cultural society such as North America, it is expected that the potential for value conflict is larger than in other societies; therefore, it is imperative that we investigate the various ways these value systems may operate in our societal institutions. We may nonetheless aspire to a global or universal bioethics, but it should not be achieved at the expense of cultural particularities. This is not to argue for cultural relativism, only to emphasize the importance of considering all particular values and perspectives in our ethical discourse and policy-making process. On a more positive note, since ethics in general, and medical ethics in particular, reveal our cultural heritage and worldviews, it provides one of the best contexts in which to examine and learn about cross-cultural differences in one's community. In turn, these new understandings of our cultural diversity will improve communication and interaction, and enrich life together in a multi-cultural setting.

Chinese Studies Program
Regent College
Vancouver, Canada

NOTES

[1] See Barker (1992), Faber-Langendoen (1992), Hansen-Flasche (1994), Lantos (1988, 1989), Miles (1994), Schneiderman, Jecker and Jonsen (1990), Veatch and Spicer (1992), and Youngner (1988, 1990), which are samples of a fairly large body of literature treating this subject.

[2] Admittedly, this strand of Confucianist teaching could have been admixed with some Taoist philosophy. Even before the Han Dynasty, a distinct demarcation of the two schools is not possible, especially in the late Warring period.

[3] *Qi* is concept held in common by all schools of philosophy in Chinese culture. *Qi,* translated as material force, in its most primordial form is believed to give rise to all visible physical matters. Each physical object in the universe has its own *qi* through which the various physical entities communicate and are connected with each other.

[4] Here I draw on Mencius' synthesis summarized in *Mencius* (1984, Bk. 2A, 6AandB, and 7A). The latter is particularly important for an understanding of the body in Confucian anthropology.

[5] Is this not part of the rationale to preserve the body of Chairman Mao Ze-dong, the founder of Chinese Communism?

[6] Shun is the legendary ancient model sageking believed to have lived around the 3rd millennium BC and who successfully built an ideal society.

[7] *The Doctrine of the Mean*, Ch.13, in Chan, p.101.

[8] *Da-tong*, literally means Grand Union, is the state of an ideal society in which love, justice, equality, and altruism are practised and in which rivalry, robbery, and poverty are absent. Enunciated by early Confucianists more than two millennia ago, this ideal state has been espoused throughout the ages until the present time by rebels and reformers alike.

[9] For a brief discussion of this as a formal subject in Confucian philosophy, see *Hsün Tzu*, 'On the Rectification of Names', in Chan, 1963, pp.124-128.

REFERENCES

American Thoracic Society: 1991, 'Withholding and withdrawing life-sustaining therapy,' *American Review of Respiratory Disease*, 144(3), pp. 478-485.

Angell, M.: 1991, 'Case of Helga Wanglie – A new kind of right to die,' *New England Journal of Medicine*, 325 (7), pp. 511-512.

Barker, J.: 1992, 'Cultural diversity changing the context of medical practice,' *Western Journal of Medicine*, 157, pp. 248-254.

Beauchamp, T. L. and Childress, J. F.: 1994, *Principles of Biomedical Ethics*, 4th edition, Oxford University Press, New York.

Bedell, S. E., Delbanco, T. L., Cook, E. F., and Epstein, F. H.: 1983, 'Survival after cardio-pulmonary resuscitation in the hospital,' *New England Journal of Medicine*, 309 (10), pp. 569-576.

Blackhall, L. J.: 1987, 'Must we always use CPR?' *New England Journal of Medicine*, 317, pp. 1282-1285.

Chan, W. T.: 1963, *A Source Book in Chinese Philosophy*, Princeton University Press, Princeton, N. J.

Choron, J.: 1963, *Death and Western Thought*, MacMillan Co., New York.
Confucius: 1963, *The Analects*, D. C. Lau, (trans.), The Chinese University Press, Hong Kong, Shatin.
Council on Ethical and Judicial Affair, American Medical Association: 1991, 'Guidelines for the appropriate use of do-not-resuscitate orders,' *Journal of the American Medical Association*, 265, pp. 1868-1871.
Crigger, B. J. (ed.): 1990, 'The court and Nancy Cruzan,' *Hastings Center Report* 20 (Jan/Feb), 38 -50.
Crisp, R.: 1994, 'Quality of life and health care,' in *Medicine and Moral Reasoning*, K.W.M. Fulford, G. Gillett and J. M. Soskice, (eds.), Cambridge University Press, Cambridge.
Duke Seang: 24, *The Ch'un Ts'ew with The Tso Chuen*, Book IX, James Legge, (trans.), in *The Chinese Classic*, Vol. V, 1994, 1991, SMC Publishing Inc., Taipei.
Faber-Langendoen, K.: 1992, 'Medical futility: values, goals, and certainty,' *Journal of Laboratory and Clinical Medicine*, 120 (6), pp. 831-835.
Hammond, J. and Ward, C. G.: 1989, 'Decisions not to treat: "Do not resuscitate" orders for the burn patient in the acute setting,' *Critical Care Medicine,* 17 (2), pp. 136-138.
Hansen-Flasche, J. H.: 1991, 'When life support is futile,' *Chest*, 100 (5) pp. 1191-1192.
Hsün Tze, *Hsün Tze* 7:9, 'On Kingly Government.'
Huang Ti: 1970, *Huang Ti Nei Ching Su Wen* (The Yellow Emperor's Classic of Internal Medicine), Ilza Veith, (trans.), University of California Press, Berkeley.
Jecker, N. S.: 1994, 'Calling it quits: Stopping futile treatment and caring for patients,' *The Journal of Clinical Ethics*, Summer, pp. 138-142.
Johnson, J. T.: 1978, 'History of Protestant medical ethics,' in *Encyclopedia of Bioethics*, Vol. 3, W. T. Reich, (ed.), The Free Press, New York.
Kleinman, A. *et al.*: 1975, *Medicine in Chinese Cultures*, U. S. Department of Health, Education, and Welfare.
Kunstadter, P.: 1980, 'Medical ethics in cross-cultural and multi-cultural perspective,' *Social Science and Medicine*, 14B, pp. 289-296.
Lantos, J. D. *et al.*: 1988, 'Survival after cardiopulmonary resuscitation in babies of very low birthweight. Is CPR futile therapy?' *New England Journal of Medicine*, 318 (2), pp. 91-95.
Lantos, J. D., *et al.*: 1989, 'The illusion of futility in clinical practice,' *American Journal of Medicine*, 87, pp. 81-84.
Loewy E. and Carlson, R.: 1993, 'Futility and its wider implications: A concept in need of further examination,' *Archives of Internal Medicine*, 153, pp. 429-431.
Ma, Boying: 1994, *A History of Medicine in Chinese Culture*, People's Publishing House, Shanghai.
Mencius: 1984, *Mencius*, D. C. Lau, (trans.), The Chinese University Press, Hong Kong, Shatin.
Miles, S. H.: 1991, 'Informed demand for "non-beneficial" medical treatment,' *New England Journal of Medicine*, 325 (7), pp. 512-515.
Miles, S. H.: 1994, 'Medical futility' in *Health Care Ethics Critical Issues*, J. Monagle and D. Thomasma (eds.), Aspen Publishers, MD, Gaithersburg.
Müller, F. M.: 1966, 'The Li ki,' in *The Sacred Book of the East*, Vol. XXVII, J. Legge (trans.), Motilal Banarsidass, Delhi.
Murphy, D. J.: 1988, 'Do-not-resuscitate orders: Time for reappraisal in long-term-care Institutions,' *Journal of the American Medical Association*, 260, pp. 2098-2101.
Polanyi, M.: 1962, *Personal Knowledge: Towards a Post-Critical Philosophy*, 2nd Edition, Routledge, London.

President's Commission for the Study of Ethical Problems in Medicine and Biomedical and Behavioral Research: *Deciding to Forego Life-Sustaining Treatment* (Washington, D.C.: Government Printing Office, 1983; GPO pub no. 0383515/8673).

Schneiderman, L. J. and Jecker, N. S.: 1995, *Wrong Medicine*, The Johns Hopkins Press, Batlimore and London.

Schneiderman, L. J., Jecker, N. S., and Jonsen, A. R.: 1990, 'Medical futility: Its meaning and ethical implications,' *Annals of Internal Medicine*, 112, pp. 949-954.

Task Force on Ethics of the Society of Critical Care Medicine: 1990, 'Consensus report on the ethics of forgoing life-sustaining treatments in the critically ill,' *Critical Care Medicine*, 18(12), pp. 1435-1439.

The Book of Filial Piety (*Xiao Jing*), in *The Sacred Books of the East*, Vol. III, 1899, James Legge, (trans.) Max Müller, (ed.), Clarendon Press, Oxford.

The Book of Rites, (*Li Ki*), in *The Sacred Books of the East,* Vol. XXVII, 1966, F. Max Müller, (ed.) Motilal Banarsidass, Delhi.

Tomlinson, T. and Brody, H.: 1990, 'Futility and the ethics of resuscitation,' *Journal of the American Medical Association*, 264 (10), pp. 1276-1280.

Truog, R. D., Brett, A. S., and Frader, J.: 1992, 'The problem with futility,' *New England Journal of Medicine*, 326, pp. 1560-1564.

Tu, Wei-ming: 1979, *Humanity and Self-cultivation: Essays in Confucian Thought*, Asian Humanities Press, Berkeley.

Veatch, R. M. and Spicer, C. M.: 1992, 'Medically futile care: The role of the physician in setting limits,' *American Journal of Law and Medicine,* 18 (1and2), pp. 15-36.

Waisel, D. B. and Truog, R. D.: 1995, 'The cardiopulmonary resuscitation not-indicated order: Futility revisited,' *Annals of Internal Medicine*, 122 (4), pp. 304-308.

Watson, J. L. and Rawski, E. S. (eds.): 1988, *Death Ritual in Late Imperial and Modern China,* SMC Publishing Inc., Taipei. Quote of David Crockett Graham, *Folk Religion in Southwest China*.

Youngner, S. J.: 1988, 'Who defines futility?' *Journal of the American Medical Association*, 264, pp. 20942095.

Youngner, S. J.: 1990, 'Futility in context,' *Journal of the American Medical Association*, 264 (10), pp. 1295-1296.

PART III

"HUMAN DRUGS" AND HUMAN EXPERIMENTATION

JING-BAO NIE

"HUMAN DRUGS" IN CHINESE MEDICINE AND THE CONFUCIAN VIEW: AN INTERPRETIVE STUDY

I. INTRODUCTION

In the name (or for the purpose) of healing and curing, human bodies – living and dead – have been manipulated in a variety of ways since the beginning of human history. They have been sacrificed, fasted, cleaned, gazed upon, touched, examined, operated on, frozen, cut, dissected, displayed, and experimented on in medical practice, research, and teaching. Human biological materials have even been directly used for therapeutic ends. For instance, in contemporary biomedicine, human blood is transferred and organs and tissues are transplanted from one body to the another. In traditional Chinese pharmaceutics, some parts, excreta, and appendages of the human body have been employed as medicine.

This chapter will offer a historical-ethical study of the Chinese knowledge and practice of "human drugs;" i.e. drugs derived from the human body. About three dozen human drugs, which include hair, fingernails, placenta, urine, bone, flesh, blood, menstrual blood, seminal fluid, and the penis, have been recorded in Chinese materia medica. The brilliant sixteenth-century physician-scholar Li Shi-Zhen systematically summarized them in a special section of his monumental work *Bencao Gangmu* (The Great Pharmacopeias). According to traditional Chinese pharmacological literature, even pubic hair was claimed to be therapeutic, with power to cure snakebite, difficult birth, abnormal urination, the yin-yang exchange disorder (a disease entity peculiar to Chinese medicine based on its understanding of sexual behaviors to health), and the ox that suffered from bloating. In order to provide a primary picture about human drugs – the diseases they treat, their therapeutic effects, and the methods of taking this kind of medicine, let me cite the whole entry "y*inmao*" (literally, shady hair; i.e., pubes) from the *Bencao Gangmu*:

> PUBES [First recorded as medicine by] *Shiyi* [i.e., *Bencao Shiyi* (A Supplement to the Herbal) of Cheng Chang-Qi, an eighth-century physician]

Ruiping Fan (ed.), Confucian Bioethics, 167–206.

[MAIN THERAPEUTIC USES] The male pubes treat snakebite. Keeping twenty pieces in mouth and swallowing with juice can prevent the poison (of snake) from entering into the abdomen. [From Cheng] Chang-Qi ['s *Bencao Shiyi*]. For the difficult birth, burn and grind fourteen pieces of pubic hairs of her husband, make the pill with lard as large as soybean and swallow it. [From Sun Si-Miao's] *Qianjing Fang* [(Prescriptions Worth a Thousand Gold)]. Female pubic hair treat five types of abnormal urination and the disease of yin-yang exchange. [Li] Shi-Zhen [first gathered].
[SUPPLEMENTARY PRESCRIPTIONS]: The Disease of Yin-yang Exchange: If, because of having sexual intercourse when one has just recovered from illness, each or both of the testicles is swollen or drawn back into the abdomen and the patient is suffering angina to death, take the female pubes, burn them to ash, and drink the ashes with the water that have washed the pudenda. [From the] *Shengji Zonglu* [(Imperial Encyclopedia of Medicine)]. The Ox Bloating to Death: This can be cured immediately by feeding the ox with female public hair wrapped in grass. [From Wang Tao's] *Waitai Miyao* [(Medical Secrets of an Official)] (Li, Vol 52: Pubes).[1]

For modern readers in the West as well as in China, the above quotation is not only exotic but also nonsensical, if not shocking. I laughed when I first read it while preparing this work even though I am trained in Chinese medicine and its history. Soon I started to be deeply puzzled: How could such a great physician-scholar as Li Shi-Zhen, along with other ancient Chinese physicians and scholars, have written this kind of nonsense? How should we approach the knowledge and practice of what Chinese physicians called human drugs? What were the ethical responses, if any, of ancient physicians and scholars toward human drugs? Is there any hidden historical and cultural information or meaning? If yes, what is this information or meaning? More importantly, is this information or meaning relevant to our time?

In the following discussion, I will attempt to give some answers to these questions. Among the many interesting issues regarding the practice of human drugs in traditional Chinese medicine, this study focuses on the historical, ethical, and cultural aspects. I will first present my methodological orientation – an interpretative or hermeneutical approach which is distinctive from the extant "iconoclastic," "progressivist," and "postmodernist" approaches. Using Li Shi-Zhen's work as the primary historical material, I will then explore human drugs in Chinese pharmacology from the context of Chinese culture. Moreover, I will explore the Confucian physician Li Shi-Zhen's moral attitude toward the subject. Finally, since the Chinese practice is not only still partly viable but also ethically relevant to the new technology of organ transplantation, I will point out the ethical significance this historical analysis of human drugs and the Confucian perspective has for today. In other words, this study attempts to bring light to the historical-cultural complexities of human drugs in Chinese society, and to illustrate the necessity and importance of developing a Confucian framework of the medical uses of human biological materials.

II. METHODOLOGY: A HERMENEUTICAL OR INTERPRETIVE APPROACH

The knowledge of human drugs, however inconceivable to us, seemed constitute an organic part of ancient common learning of curing and healing. There are many exotic claims, "apparent absurdities," or myths in the Chinese knowledge of human drugs, as indicated well in the passage quoted above concerning the medical effects of pubic hair. If the passage came from a work by an odd and abnormal ancient author, we might take the quotation lightly or just ignore it. But the *Bencao Gangmu* is the greatest pharmacological work in the entire history of pre-modern China and still a standard reference for the subject. In the words of the historian of Chinese medicine Paul Unschuld (1986, pp. 145, 163), it "is the best-known and most respected description of traditional pharmaceutics" and "the qualitative and quantitative climax" in the development of materia medica literature. The author and compiler Li Shi-Zhen is widely considered to be one of the world-class ancient Chinese scientists and physicians. Moreover, all four works cited by Li in the entry "Pubes" – the *Bencao Shiyi*, the *Qianjing Fang*, the *Shengji Zongu*, and the *Waitai Miyao*, especially the latter three, belong to the well-known and highly-respected classics in the history of Chinese medicine. The seventh-century scholar-physician Sun Si-Miao, the author of the *Qianjing Fang*, was so prominent that he was called the "king of medicine." Therefore, although it is impossible to know exactly what percentage of ancient physicians, scholars and the common people believed in the therapeutic effects of pubic hair in particular and human drugs in general, at least for the authors who wrote about the pubes as medicine, the above quotation represented some real and significant knowledge no matter how strange, unintelligible, or laughable its content sounds to us.

Thus, we are placed in a puzzling situation similar to what Thomas Kuhn faced when he read Aristotle's theories of natural phenomena, especially those about the motion of the objects, which are obviously nonsensical from the standpoint of modern physics. Kuhn later reflected upon his experience in this way: "When reading the works of an important thinker, look first for the apparent absurdities in the text and ask yourself how a sensible person could have written them" (1977, p. xii). For him, these absurdities in the texts of the history of science can be read and interpreted in quite different ways. He went on to point out that: "When you find an answer, ... when those passages make sense, then you may find that more central passages, ones you previously thought you understood, have changed meaning" (Ibid.).

How to read and interpret the apparent absurdities which widely exist not only in the historical literature of medicine and science also the other cultures or life-worlds? In this century, at least three different approaches have already been developed. I would like to call them respectively "iconoclastic," "progressivist," and "postmodernist." These different ways of reading reflect not only the interpreter's understanding of the apparent absurdities in traditional or foreign texts but also his or her general evaluation of tradition, history, or foreign culture.

The iconoclastic approach sees these apparent absurdities as absolute nonsense, unscientific and superstitious stuff that came down from our past which is often viewed as quite dark. In China, the typical holders of this view were the intellectual and medical iconoclasts in the early decades of this century. In the May Fourth Movement of New Culture, science replaced the orthodox Confucianism and became the highest and absolute standard of value. It was professed that only what is compatible with science should live and anything else should die. In the view of these iconoclasts, not only does the absurdities in the passage quoted above provide one of the most convincing proofs about the incompatibility of traditional Chinese medicine with modern science, the un- and anti-scientific knowledge in traditional medicine also stand as a representative or symbol of the unenlightened past and all that was backward, irrational, and superstitious in the old society that should be destroyed.

The dominant and standard view on traditional Chinese medicine in contemporary China represents the second approach – what I call a progressivist one. According to this point of view, Chinese medicine is a science, although in an ancient form. That is to say, the great scientific essence is mixed with the superstitious dross. On the one hand, it affirms that there is much un-scientific, if not totally anti-scientific, dross, of which the traditional understanding of the medical use of pubes can be regarded as a striking example. On the other hand, the progressivist approach rejects the notion that traditional medicine should be eradicated because of the existence of some superstitious stuff. Progressivist critics assert that there exists merely a small portion of unscientific content in Chinese medicine and that, on the whole, traditional medicine is a science with its own theoretical foundations and practical values. To traditional medicine, they believe, what is necessary is to discard the dross and select the essence. Even though people never reach a consensus about what actually constitutes the dross and what is the essence, most, if not all, agree that the use of pubes for medical purpose is that which needs to be discarded, in the least not taken seriously.

Despite the dissimilarities and disagreements between iconoclast and progressivist approaches, they share some common assumptions. For example, both assume that we as interpreters can be and actually are outside of tradition and history and that modern science is the standard of judging traditional Chinese medicine. Both of them hold that we, with our unprecedented science and technology, rationality, and advanced knowledge, are intellectually and even morally superior to the ancient physicians, scholars, and people in general. Consequently, it is we modern people who are capable of either establishing a new scientific medical system through abandoning the old tradition or differentiating between what is essence and what is dross. Obviously, all these assumptions are consistent with the modern world-view of science, truth, tradition, and history; thus, both "iconoclastic" and "progressivist" approaches can be generally called "modernist."

In contrast with the modernist views, the approach that I call postmodernist gives much favorable attention to the divergence between modern biomedicine and traditional healing systems. In this view, the apparent absurdities in historical texts dramatize the differences of various systematic structures. Actually, what Michael Foucault has said in his *The Birth of Clinic* (first published in 1963) about Western medical perceptions in different historical periods is very helpful for geting a sense of the postmodernist perspective on the fundamental distinctions between traditional Chinese and modern scientific medicines, especially the apparent absurdities of human drugs in Chinese pharmacology.

> Not only the names of diseases, not only the grouping systems were not the same; but the fundamental perceptual codes that were applied to patients' bodies, that field of objects to which observation addressed itself, the surface and depths traversed by the doctor's gaze, the whole system of orientation of his gaze also varied (Foucault, 1973, p. 54).

According to early Foucaultian thought, the material on pubes in the *Bencao Gangmu* sounds absurd to us only because we live in a discourse or systematic structure different from that of the authors of the text. This is just like the example of the middle eighteenth century physician Pomme's description of "membranous tissues" and the classification of animals in the ancient Chinese encyclopedia, which are cited at the very beginnings of *The Birth of the Clinic* and *The Order of Things* respectively, makes little sense to twentieth-century minds. Similarly, Kuhn (1970) suggested that the apparent absurdities of the great ancient scientific works are absurd only when we attempt to understand them from our own perspective. Many apparent absurdities will disappear if we approach ancient thought not as the primitive beginning of our own paradigm, but as a paradigm parallel and competing with ours. The meanings of ancient texts are systematically distorted if they are not read or interpreted as something representing a world-view or a paradigm incommensurable with that one we commonly share.

It is symbolic that Foucault later shifted his methodology from the analysis of structure or discourse to interpretive analysis (Dreyfus and Rabinow, 1983). The general theoretical and methodological orientation I assume in this study of human drugs in Chinese medical tradition and the Confucian view is hermeneutical or "interpretative." Historically, one of the primary use of hermeneutics has been its application in biblical interpretation, but more recently, the meaning and scope of hermeneutics have significantly expanded through the works of the thinkers like Schleiermacher, Dilthey, Heidegger, Gadamer and Ricoeur. Because of its unique visions regarding language, tradition, understanding, the text, the interpreter, the interpreted, practice, etc., contemporary philosophical hermeneutics has been applied to many other fields, such as anthropology (see Geertz, 1973), history (see Mink, 1991), social and cultural critics (see Walzer, 1987), and medical ethics (see Carson, 1990; ms.). Hermeneutics should not be understood as a group of methodological principles of interpretation or a method of understanding although it does mainly concern understanding, meaning, and interpretation. In his influential book *Truth and Method*, Gadamer argues, in an ironic way, that the modern obsession with Truth and Method – reaching Truth by means of Method – has distorted and concealed the ontological character of understanding. Contemporary hermeneutics never aims to invent the objective method of understanding or to discover the proper meaning and the best interpretation. This was one of the great, yet unrealized, modernism's dreams.

As Geertz, Mink, Walzer, and Carson have illustrated, an hermeneutical or interpretative approach to culture, history, social practice, medical ethics has a series of distinctive characteristics. An interpretative approach to the Chinese practice of human drugs in particular and the medical traditions in China in general puts stress on the following four points:

First, an interpretative approach is discontent with influential scientism and attempt to go beyond its heavy shadow. Scientism, a belief in which modern science has been used or even worshiped as the only standard for truth, knowing, and evaluation of all the other systems of knowledge, has dominated twentieth-century understanding and interpretation of traditional Chinese medicine (Nie, 1989; 1995). This trend of thought has also become imbued in the modern historiography of Chinese healing heritage (Nie, 1997). Despite the many apparent absurdities in employing human drugs, the twentieth-century historians of Chinese science and medicine have found that there exist some great scientific discoveries and achievements in this practice. After carefully examining the entries related to human urine in the *Bencao Gangmu* – urine, urine sediment, urinary calculi, and especially the concrete procedures of preparing "autumn stone" which was coined by a Daoist prince some time before 125 B.C.E., the eminent historian of Chinese science Joseph Needham and his chief collaborator Lu Gwei-Djien concluded that the ancient Chinese remarkably anticipated such Western discoveries and inventions as rich steroid sex hormones in the urine of pregnant women and techniques of isolating sex and pituitary hormones from human urine by centuries. They considered the work done by the ancient Chinese "a brilliant and courageous anticipation of the conscious biochemistry of our own time" (Needham and Lu, 1983, p. 336).

Due to the influence of scientism, Chinese medicine is seen either as un- and even anti-scientific or as a pre-scientific – an ancient counterpart of modern biomedicine, rather than a paradigm or discourse parallel with modern science and medicine. In the former case, the differences between Chinese medicine and sciences have been exaggerated to the extent that Chinese medicine is regarded as mere hocus-pocus. In the latter case, concepts, theories, methods, and procedures from modern sciences have been applied to study Chinese medicine and a great effort has been made to discover the consistence between and convergence of between the two systems. This view emphasizes that Chinese medicine is fundamentally compatible with modern sciences, and moreover, value that Chinese medicine is able to provide provides a sort of spiritual guide revelation or enlightenment for biomedicine owing to its rich philosophical wisdom and strong holism.

Scientism has seriously hindered the full appreciation of conceptual richness and historical-cultural complexities of Chinese medical heritage (Nie, 1995, 1997). Unquestionably, some science is embedded in ancient Chinese medicine, Chinese medicine in general and the Chinese practice of human drugs in particular. However, there is much more to the medical tradition than science and scientific achievements. For example, a lot of fundamental differences, incommensuabilities, apparent absurdities, or myths surrounds the practice of human drugs. The myths not only reflect more vividly than the scientific elements the Chinese medical professional's understandings of the human body, illness and disease, nature, the supernatural, and the art of healing. They have also played a more active and significant role than its scientific elements in the everyday Chinese practice of health care and the popular ideas about body, health, illness, and spiritual existence. It is the apparent absurdities and myths that represent a Chinese way of grasping and expressing some awe-inspiring connections between human existence, health, illness, nature, and the supernatural.

I use the word "myth" to refer either 1) a belief/notion that is legitimate and meaningful for living and thinking in particular discourse, e.g., Chinese medicine and culture, but unfounded and absurd in another discourse, e.g., modern medicine and science and Western culture, or 2)simply a basic intellectual assumption in any discourse that has not yet been seriously examined but may be essentially questionable and problematic. It should also be pointed out that a myth as a conceptual system not only is based on a certain social practice but also shapes or frames this practice. In this sense, a myth is an expression of a particular form of practice.

As a result, the second feature of interpretative approach is that it not only takes the apparent absurdities or myths very seriously but also appreciate them. Like the postmodernist approach, an interpretative orientation treasures the myths, apparent absurdities or fundamental differences. It aims at understanding and deciphering the cultural meanings of these differences, rather seeking for the best and only interpretation. Against the modernist approaches which simply turn away from or totally rebuke the apparent absurdities in the practice of human drugs and in historical texts, the hermeneutical approach never takes these myths lightly. It presumes that, however absurd, the exotic claims concerning human drugs are like drops of water from the huge ocean of Chinese medicine, culture, and society. Actually, the very short entry "pubes" in the *Bencao Gangmu* raises many interesting questions regarding Chinese views of the human body, gender difference, sexuality, the cause of disease, the magic character of number, the relationship between humankind and nature, the relationship between human beings and animals, etc.

The third characteristics of interpretative orientation is that it tries to comprehend both dissimilarities and commonalities of cultures, discourses, or practices. Unlike the strictly postmodernist approach which emphasizes the incommensurabilities in different cultures, life-worlds, discourses, or paradigms, the interpretative approach holds that a postmodernist reading and appreciation of absurdities or differences of the "other" is necessary but not sufficient because this view merely honoring the differences. To better understand the others, it is necessary to employ what the anthropologist Clifford Geertz (1983, p. 16) called "the merest decency," by which he means "to see others as sharing a nature with ourselves." The decency has often often forgotten in the modern discourse due to our arrogance to the other. A number of amazing commonalities, differences, commonalities in differences, and differences in commonalities between traditional Chinese healing and modern scientific medicine provide marvelous source materials for cross-cultural comparative studies from the various disciplines such as history, anthropology, linguistics, philosophy and sciences, and, more significantly, from interdisciplinary perspectives (see He, *et al.*, 1990). The practice of so-called "human drugs" in traditional Chinese pharmacology provides an illustration of the fundamental differences and the surprising similarities between biomedicine – today's orthodox and legitimate medical system – and Chinese healing tradition. For example, in spite of the apparent absurdities of human drugs, the practice shares a fundamental similarity with the modern technology of organ and tissue transplantation – using human biological materials for the medical purposes.

The forth feature of an interpretative approach to medicine and ethics is its contextual way of seeing and thinking – its sensitivity to and emphasis on the historical and cultural context. Both medicine and ethics always tend to be universalistic. Actually, consistent with modern discourse, the most common approach to medical ethical issues in the present seeks to analyze, justify or criticize, and judge the moral problems in the settings related to health care from one or some universal and absolute principles. Yet, the methodological approach of this essay is historical, descriptive, articulative, and interpretative. Through intensively reading and re-reading the specific volume on human drugs in Li Shi-Zhen's *Bencao Gangmu*, I will attempt to approach the Chinese practice of human drugs from both an insider's and an outsider's points of views. My ethical analysis of this practice will be mainly based on the sympathetic understanding and critical interpretation of its historical-cultural background. In section V, I will discuss further the importance of historical-cultural context in addressing medical ethical issues.

In order to appreciate both differences and similarities between Chinese and Western medicine and their historical-cultural contextual complexities, an interpretation is intrinsically multi-disciplinary and inter-disciplinary. Its literary genre is similar to what the medical ethicist and historian Albert Jonsen (1990, p. 4) calls "secular aggadah," which "move without embarrassment among myth, history, science and philosophy." In this study, I will try to move, possibly awkwardly without embarrassment, among history, philosophy, science and myth, between traditional healing and contemporary biomedicine, among the Western perspectives and Chinese visions, and between the past and the present.

III. THE PRACTICE OF HUMAN DRUGS IN HISTORICAL-CULTURAL CONTEXT

Human drugs in Chinese medicine embody rich information of how the Chinese perspectives on the body and human relationship, disease and illness, nature and the supernatural, and the art of medicine. In this section I will attempt to explicate the historical-cultural context of the knowledge and practice of human drugs in pre-modern China.

1. Human Drugs in Chinese Medicine and the Role of Daoism

Although the Chinese materia medica consists of substances from animals, including human beings, plants, and minerals, the majority of Chinese medicinals are vegetable origins, i.e., herbs. As a matter of fact, for centuries Chinese pharmacology has been conventionally termed "*bencao,*" which literally means roots and grass. The earliest systematic book of materia medica, which appeare in the second century and is attributed to the ancient emperor Sheng Nong ("Divine Peasant" or "Legendary Farmer"), is entitled *Shengnong Bencao Jing* (The Divine Peasant's Classic on Materia Medica). Almost all important subsequent pharmaceutical works has been titled "*bencao.*" Nevertheless, the human drug has been used by Chinese physicians and included in literature of Chinese pharmacology since the second century. In the *Shengnong Bencao*, one substance originating in the human being was recorded. This substance is human hair and it belonged to the superior class of drugs.[2]

Up to the sixteenth century, thirty-five human drugs had been recorded in the great pharmacological work *Bencao Gangmu*,[3] published first in 1597 soon after Li Shi-Zhen died. It is generally considered the most systematic summary of all materia medica literature in traditional China. This generalization regarding the position of the *Bencao Gangmu* in the history of Chinese pharmacology is especially valid as far as human drugs are concerned. Using the principle of arraying from the small to the big and from the low to the noble, Li divided nearly seventeen-hundred medical substances gathered in his work into sixteen sections: water, fire, earth, metal and stone, grass, cereal vegetable, fruit, tree, clothes and utensils, insect, creature with scales (reptiles, fish), creatures with shells, birds, quadrupeds, and finally, materials with human origins. He dedicated the whole fifty-second volume, i.e., the very last one, to human drugs.

The thirty-five agents in the "human section" of *Bencao Gangmu* are: hair, hair from the comb, dandruff, earwax, dirt from the kneecap, fingernails, teeth, excrement (and licorice in human feces, i.e., powdered licorice root which has been enclosed in a bamboo case and buried in a cesspool for one winter, the case being hung to dry thoroughly, and the licorice extracted), fetal urine of the infant (the very first urine of the new-born baby?), urine (especially the boy's urine), urine sediment (the dried white precipitate found in urinary chamber pots), "autumn stone" (purified and concentrated urine preparations), urinary calculi, proclivity stones (the stones are formed within his/her body because someone is so absorbed in something), milk, menses (and sanitary towel), blood, male semen and female seminal fluid, saliva, soft deposit scraped from teeth, perspiration, tears, breath, the soul of the human (a material object resembling pine charcoal dug out of the ground beneath the body shortly after a hanged person died), facial hair (mustache, beard, and sideburns), pubes, bones, the bregma, placenta, the water of placenta (liquid from a placenta that has decomposed after long burial), umbilical cord, penis, gall, flesh, and the mummy (a "honey man," i.e., the body of a old man, who drank only honey during his last days, became honey after having being buried in honey for a hundred years).

The Chinese use of human drugs is closely related to Daoism (an indigenous philosophical-political-religious system second only to Confucianism in importance and influence), and particularly, to the Daoist understandings of and attitudes toward the human body and the place of human beings in nature. As the monumental volumes in the general title *Science and Civilization in China* by Needham and his collaborators have shown, we owe to the Daoist the "beginnings of chemistry, mineralogy, botany, zoology and pharmaceutics." Daoism is of such "cardinal importance" for the Chinese history of science that "China without Daoism would be a tree of which some of its deepest roots had perished" (Needham and Wang, 1956, pp. 161-162). This is also the case with regard to human drugs. Some human drugs were first invented or discovered by Daoist adepts, who were traditionally called "*fangshi*" (magicians). Some became popular because of efforts of the believers of Daoist religion.

There exist many other names or euphemisms for human drugs. These names reveal not only the direct and indirect historical connections of the origins of human drugs in Daoism, especially the Daoist religion, but Daoist mythical understandings of human body parts. For human placenta, there are "the clothing of the fetus," "the purple chariot of river," "the clothing of the Chaos," "the very first mother of the Chaos," and "the clothing of the immortal," (interestingly, another is "kasaya of the Buddha"); for urine, "the soup of returning to the Origin" and "the wine of transmigration"; for excrement, "the soup of the yellow dragon" and "the water of returning the Origin;" for milk, "the wine of the immortal"; for saliva, "the spiritual juice," "the divine water," "the golden thick liquid," and "sweet wine"; for menstrual blood, "the red elixir." The terms like "returning to the Origin," "the gold elixir," "the immortal," and "the Chaos" belong almost exclusively to the Daoist dictionary.

The medical use of salvia, milk, and seminal fluid are highly related to the theories and practice of what the Daoists called the art of bedchamber – using human sexuality as a technique to achieve longevity and even physical immortality. There was a Daoist doctrine "the great medicine of the three peaks" in which the medicinal properties of such female secretions as salvia, milk, and seminal fluid were formulated and promoted (see Gulik 1961, pp. 282-234).

Nevertheless, it is inaccurate to attribute the historical origins of all human drugs to Daoist adepts. Daoism has played a significant role in shaping the Chinese practice, but it is not the only force. Actually, most human drugs such as hair, dandruff, earwax, dirt from the kneecap, fingernails, teeth, facial hair, pubic hair, gall, and flesh are hardly seen as the Daoist inventions. There exist other factors like Chinese medical thought and some collective cultural beliefs of Confucianism which have resulted in the practice of human drugs jointly with Daoism.

2. The Efficacy of Human Drugs

Human drugs had been applied to various health problems – physical, psychological, and supernatural. I quoted the health problems to which pubic hair can be employed. Let me take hair, finger-nails, and menstrual blood as further examples to show the symptoms, illnesses, and diseases claimed by Chinese physicians to be curable by human drugs alone or together with other medicinal materials. According to the literature of ancient materia medica collected in *Bencao Gangmu*, it is claimed that human hair can treat difficulties of urination, fright of the child and the adult, incised wound, bleeding, wind damages, blood dysentery, painful and rough urination caused by stones of the urinary bladder, yellow disease due to cold factor, un-expelled placenta, possession of a small child by an offended spirit, quick stomach with aching, and felons and suppurating abscesses. The disorders that could be cured by nails include: three Corpse-Worms (within the human body), foot-*qi* (beri-beri), open wound entered by wind, Yin-yang exchange disorders, swollen belly in infants, abnormal urination in men and women, hematuria, hematuria in pregnant women, un-expelled placenta, painful ulcerated hemorrhoids, metal needles stuck in the flesh, flying gossamer in the eye, cataract due to smallpox, eye diseases in general, chronic bloody diarrhea, and nosebleed. Menstrual blood could be employed against relapse caused by fatigue in women, recurrence of a hot disease due to fatigue, yellow jaundice due to fatigue in women, acute sudden delirium, convulsions in children due to fright, jealousy of the wife (the woman), carbuncles on the back, sores on the penis, arrow poisoning, arrowhead lodged in the abdomen, horse's blood entering a sore or a wound piercing the skin while skinning a horse, and sore wounds from tigers and wolves.

But are human drugs clinically effective? Few contemporary people in the West as well as in China would believe in the mentioned therapeutical effects of pubes, hair, finger-nails, and menstrual blood without strong doubt. Actually, more than twenty years ago, medical scientist William Cooper and historian of Chinese science Nathan Sivin analyzed eight representative human drugs – hair, nails, teeth, milk, blood (blood in general and menstrual blood), male and female seminal fluid, saliva, and bone (and the bregma) – from the perspective of scientific medicine. They discovered that less than eight percent of all the symptoms and diseases mentioned in the part of "supplementary prescriptions" could be benefitted by the main and ancillary ingredients of the remedies cited. If considering only the effects of the main ingredients of prescriptions, the human drugs could positively effect fewer than five percent of the disease entities or symptoms (Cooper and Sivin, 1973, pp. 259-261).

However, clinical efficacy is anything but a mere scientific or statistical issue. A number of social, historical, religious, cultural, and personal factors involved in evaluating whether and how much a drug or therapeutic measure is effective. In his classic work of medical anthropology *Patient and Healers in the Context of Culture*, Arthur Kleinman regards "how we evaluate therapeutic efficacy as the central problem in the cross-cultural study of healing." (1980, p. 314) Based on the empirical data from his fieldwork in Taiwan, he clearly demonstrates that, without taking the complex cultural factors into account, it is impossible to discuss the issue of clinical efficacy of health care, especially indigenous healing – whether religious or secular.

It is obvious that Li Shi-Zhen and many other traditional physicians believed in the efficacy of human drugs in the treatment of most health problems recorded in literature. For ancient Chinese the knowledge of employing human drugs was both understandable in theory and valuable in practice. In their historical and scientific study of human drugs in

Chinese medicine – so far the most systematic one on the subject in the West as well as in China – Cooper and Sivin have focused on the "interplay of empirical, psychosomatic, ritual, and theoretical factors in the Chinese doctor's evaluation of his clinical experience." Through seriously examining "the process by which certain methods and agents without demonstrable pharmaco-dynamic efficacy under the conditions of modern medical experimentation were retained in use by Chinese physicians," they found out that "there is no such thing in Chinese drug therapy as an empirical tradition based on pharmacological properties alon. Magic and ritual play too large a role in drug formulas to ignore, and clinical experience was always shaped and interpreted in terms of theory"(1973, p. 206). In other words, one is never able to understand the issue regarding the efficacy of any medicine or healing method in Chinese society including human drugs without properly appreciating the Chinese visions of various diseases and illness, the causes of health disorders, and the historical-cultural context.

3. Placenta and the Chinese Illness Xulao *(General Weakness)*

While most human drugs have been abandoned in contemporary China, some such as placenta, nails, child's urine, ashen hair, licorice in human feces, and urinary sediment are still used and included in recent textbooks and handbooks on traditional Chinese pharmacology. Among them, the most widely employed is the human placenta. There are "injection fluid of placenta" and other preparations of medicinals made from placenta in the pharmacies. Human placenta is generally classified in the category of *yin* and blood tonics – drugs for nourishing the blood and *yin* essence. In a dictionary of traditional Chinese medicine, it reads: "The drug consists of the dried human placenta. It is used to replenish the vital energy, to nourish the blood and essence for general weakness, anemia and neurasthenia" (Xie and Huang, 1984, p. 208). In one of the appendices of a contemporary work on Chinese herbs which was written by a Western scholar and published in Japan in the English language, it tells that dried human placenta can be apply to neurasthenia, impotence, infecundity, and pulmonary tuberculosis (Keys 1976, p. 315).

Among many other names for placenta, the most common in use is "the purple chariot of river," which comes from the Daoist understanding that the fetus in the placenta is like driving a chariot in the water of the womb and that the placenta with the purple color is finer than red or green. It is believed that the placenta of the woman's first birth is the best. Second best is that placenta from the birth of any healthy woman. The gender difference of afterbirth is noted, but no commonly accepted theory about this difference has been established[4] (Li: Vol. 52, Human Placenta).

According to the historical materials presented by Li Shi-Zhen, the medical use of human placenta originated in a country neighboring China, which the ancient Chinese considered uncivilized. The official history of the Sui Dynasty recorded that the woman in Liu Qiu always eats the 'clothe of the fetus' when she has given birth to a baby. Although it had been gathered in the *Bencao Shiyi* of the eighth-century pharmacologist Cheng Chang-Qi, placenta as a tonic drug did not become popular until two events occurred. First, the well-known fourteenth-century physician Zhu Zhen-Heng claimed its medical effects for consumption, and second, the doctor Wu Qiu, one of Zhu's contemporaries, invented "the Great Creation Bolus" in which the placenta was combined with many other Chinese materia medica.

The popularity of "the purple chariot of river" was greatly increased with the invention of the Great Creation Bolus. Li Shi-Zhen quoted in length what the inventor Wu Qiu said about human placenta, the "magic" effects of the bolus, and the reasons for these effects.

> The child is bred in the womb. The umbilical cord is connected to the placenta and the placenta to the spine of the mother. Benefitting from the mother's privilege, the placenta is formed by the essence from the father and the blood from the mother. It is cherished by the genuine vitality. Therefore, "the chariot of river" inherits indeed the *qi* of earlier heaven [inborn vital energy] although it has the form of later heaven [acquired form] only. As a result, its medical effects excels those of the other drugs originated in metals, stones, grass and trees. I always get success in employing it. It is even more effective to women. This is because the placenta comes from the woman and both are of a kind. If one suffers from infertility, giving birth to more girl babies [less boy babies], menstrual irregularities, miscarriage, and difficult delivery, the placenta will definitely make her to have sons. Someone dying from the fatal disease can live one or two days longer by taking of one or two doses [of the Great Creation Bolus]. The bolus is especially good at supplementing the yin essence as one hundred shots hits one hundred targets. If taking enduringly, one will be able to see and hear clearly, have glossy black hair on the head and at the temples,

> prolong one's life, and carry off the functions of the Creator. Hereby it is named as "the Great Creation Bolus" (Li: Vol. 52, Human Placenta).

After expounding the complex technical process of preparing the bolus and its variations in detail, Wu Qiu further reports four concrete cases to support his claim about its miraculous results. They are the following: a weak and impotent patient changed his physique and complexion almost totally and begot four children in succession after taking two doses; a decrepit sixty-year-old woman lived to ninety years old and remained strong and healthy because of the bolus; a man who could not speak after a disease recovered a sonorous voice due to the medicine; and a patient, who had suffered from wilting feet and could not walk for half a year, was able to walk a long distance after taking the bolus (Ibid.).

The main "disease" that dried human placenta treats is *xulao* (general weakness and exhaustion), also named *xu* (deficiency, depletion), *xusun* (vacuity detriment), or *xu ruo* (vacuity, weakness). From the medical perspective, general symptoms or signs of *xulao* include: lethargy; languor; frail and weak movement; ashen, pale, or sallow face; regular throbbing of the heart and shortness of breath; pain that is relieved by pressure; perspiration without external cause and nocturnal sweating; and copious urination or incontinence. Medical examination finds such signs as soft tongue and pale tongue material with little or no moss, and small, empty, thin, weak and debilitated pulse. It is impossible to name a corresponding disease entity in Western medicine even though *xu* or *xulao* seems to be highly compatible with neurasthenia – weakness of nerves and nervous exhaustion. Despite its past vogue in the West, the disease name neurasthenia has been displaced by the terms like "depression" and "stress" nowadays. But it is still widely used in China. Kleinman notes that one of the reasons for the popularity of neurasthenia in China is that "neurasthenia as a concept fairly easy to assimilate to traditional Chinese medicine, which has since ancient times had an interest in problems such as weakness and fatigue" (1988, p. 109). Nevertheless, even though both *xulao* and neurasthenia refer to general weakness, they are not identical disease entities. While both include bodily and mental complaints and signs, neurasthenia has more emotional symptoms with its neurological origin and *xulao* is more somatic, more concerned with the weakness of constitution.

Just as neurasthenia is shaped by and diffused into Victorian American culture, *xulao* can be seen as the Chinese illness. It has great medical significance for traditional style physicians. The Pattern of Deficiency is one of the eight principal patterns, and the replenishing or reenforcing method is one of the most fundamental methods of treatment. As a general term for the methods of treating various deficiency symptom-complexes with tonics, the replenishing method employs many drugs and prescriptions among which are human placenta and the Bolus of Great Creation. The concept of *xu* or *xulao* has also great, multifaceted social significance and many cultural meanings. For the Chinese, it is a real disease or unhealthy condition, which has been used to explain not only bodily and emotional disorders but also many social and interpersonal problems. Even someone's failures in performing certain social functions can be excused because he or she suffers from *tixu* (weak constitution). As a matter of fact, the concept of *xu* has so well woven into the ordinary Chinese social life that many tonic drugs have been widely used by lay people without requiring any medical consultation or prescription from a doctor. Among them are ginseng, *tangkuei* (Angelicae Sinensis Radix), velvet deerhorn (horn of the stag), human placenta, and even the flesh of dog (another shocking thing to the Westerners). It is not uncommon that some Chinese parents cook placenta in particular ways and let their children take it for building up their health even though their children may not be really weak at all.

4. Human Flesh as Medicine and the Confucian Conception of Xiao *(Filial Piety)*

The most intriguing in the practice of human drugs is probably the belief that certain human relationships, like that between parents and children, and certain human virtue, like *xiao* (filial piety), are indispensable in employing such human drugs as human flesh. In the Chinese view, it is not the human drug *per se* but certain human relationships and virtues that really cure.

Human flesh as an entry in the literature of materia medica first appeared in the Tang Dynasty. In the eighth-century doctor Cheng Chang-Qi's *Bencao Shiyi,* it was claimed that human flesh can treat *zhai* (consumption, chronic consumption). Recent historical studies by the Chinese historian of medicine Ma Bo-Yin (1993, p. 445) show that using human flesh as medicine seems to have originated in India. The earliest record on the medical effect of human flesh actually comes from the Buddhist classic *Miao-Yin Baozhuan* (The Treasured Volume of Ms. Miao-Yin) in which it is said that the princess Miao cut her flesh and even eyes to cure her father.[5] Partly, if not mainly, due to the emphasis of Confucianism on filial piety, cutting one's flesh to cure diseases of one's parents became quite popular in China following the Tang Dynasty. In the official history of the Tang Dynasty, four cases were found; in that of the Song Dynasty, more than ten cases; in that of the Yuan Dynasty, eight; in that of the Ming Dynasty, quite a few also (Ma, 1993, p. 446). It never completely stopped despite efforts by some emperors to prohibit the practice. In present China it was once reported that a man, when young, cut his flesh with the hope of curing his parents' chronic disease. Yet, this action is far from popular, or even acceptable nowadays.

Here it is necessary to emphasize that the human relationship between the parents and the children is the most significant factor in the use of human flesh as medicine. That is to say, it is not flesh itself but a particular human virtue, filial piety in this case, that can help the medicine prescribed by the physician to cure disease. As a result, a son or a daughter can use their own flesh to cure their parents, but a parent's flesh can never be used to cure their children. For healing one's parents, a stranger's flesh is useless.

In a sense, the "strange" practice of cutting one's flesh to treat one's parent was the result of the Chinese emphasis on the virtue of filial piety. There are many common or compatible virtues in Chinese and Western cultures such as love, wisdom, knowledge, self-examination, courage, prudence, the mean, loyalty, sincerity, altruism, righteousness, and humbleness, although these virtues may be interpreted very differently in the two cultures. However, there exist some very different or even unique virtues and moral concepts in China. The conception of filial piety is a salient example. No other philosophical-religious-political system, not even Daoism, has given such priority to the virtue of filial piety as Confucianism. The first sentence of the *Xiao Jing* (The Classic of Filial Piety) states: "Filial piety is the basis of virtue and the source of culture." According to Confucianism, filial piety is far more than merely a domestic virtue. It has developed into a general virtue in almost all aspects of life – legal, political, economical, moral, and social. Consequently, it is not surprising that filial piety is believed to have healing and curing power in China. Not only human flesh, even the shirt of the dutiful son or the son in mourning could be used as medicine. According to Li Shi-Zhen, the shirt treats the sore on the nose and the forehead by washing it secretly. Li stressed that secrecy – "Do not let people know!" – is important for the efficacy of the shirt (Li: Vol. 38, The Shirt of Mourning Son).

However, the correlation between the historical origin of human flesh as medicine and the Confucian concept of filial piety does not mean that Confucianism always ethically approves the use of human dugs. Rather, Confucianism – the legitimate ideology in pre-modern China and a still very influential Chinese system of thought – has a complex and ambivalent view of the practice, which I should now describe in some details.

IV. THE CONFUCIAN PHYSICIAN'S CRITICISM AND AMBIVALENCE

The Confucian physician-scholar Li Shi-Zhen had many reservations concerning the practice of human drugs and condemned the use of some of them even though he had contributed thirteen new human drug entries to Chinese materia medica, much more than anyone else has done. Just as no medical thought or practice can be free from values and no physician can be free from the social and cultural influence in which he or she lives, Li's attitude towards human drugs was shaped by his Confucian beliefs. Beyond any doubt, Li Shi-Zhen was a "*ruyi*" (Confucian physician and medical scholar) in the full sense of the term. At least three facts support this statement: first, his anti-Daoism position; second, his commitment to Confucianism, especially to Neo-Confucianism; and third and most important, his ethical analysis and criticism of the use of the human drug from Confucian standpoint.

Daoism and Confucianism constituted the two competing, though sometimes complementary, political-philosophical-religious systems in traditional China. Although most scholars and common people did not believe exclusively in Confucianism, Daoism, or Buddhism, many Confucian scholars often opposed the other two. As a Confucian physician, Li Shi-Zhen often took the fact of the Daoist origins of human drugs as a reason to reject them. Concerning the Daoist practice of taking the male and female seminal fluid, Li harshly criticized:

> Black magician bewitched and poisoned the foolish people. They took the virgin girl to have sexual intercourse and drank her seminal fluid. Or they drank and ate their own semen with menstrual blood. They even called this "lead and mercury" [the materials of longevity and

> immorality according to Daoist alchemy] and considered it "secret prescriptions." Their natural years were shortened by absorbing in greedy and dissoluteness and by eating the filthy and dregs. Oh, this is the most stupid. And who should be accused? (Li: Vol. 52, Human Seminal Fluid)

The Daoist adepts called human breast milk, especially that from virgin girls, "the wine of immortals," "the blood of the living person," and "the white elixir." Li reproached:

> Black magicians took virgin girls and then made milk by rubbing. There was also the idea that make menses upward to change into milk. These are nothing cheating the greedy and silly people by concocting all sorts of names. All the activities are what the sorcerer (evil person) did and the emperor's law prohibited. "*Junzi*" (the gentleman) must rebuke (Li: Vol. 52, Human Milk).

Yet, for Li Shi-Zhen the Daoist origin is not the only, or even not an important, reason for him to oppose the use and abuse of human drugs. Actually, Li accepted some human drugs with clearly Daoist origins without raising any ethical question. For instance, he confirmed the medical effects of the sediments and natural precipitates of the urine in prolonging life and curing diseases, especially sexual debility and related disorders. Moreover, he introduced the concrete methods of urinary preparation as described in the Daoist texts. The reason seems to be simple: it is almost impossible to discuss Chinese pharmacology in particular and Chinese sciences in general without referring to the Daoist activities and accomplishments.

In contrast with one of his forerunners Tao Hong-Jing (452-536 C.E.), the great Daoist who specialized in the study of materia medica, Li Shi-Zhen clearly affiliated himself with Confucianism, more exactly, Neo-Confucianism – the legitimate school of moral and political thought which, during his lifetime, had established for several centuries.[6] He attempted to become a civil official by passing the official exams, which focused exclusively on the doctrines and arts of Confucianism. He gave up the effort after failing three times and devoted himself solely to medicine that, in China, was seen as the "little Dao" inferior to the "great Dao" – a career of the official. However, Li did not give up his Neo-Confucianism and he became a Confucian physician finally. As Unschuld points out, Li Shi-Zhen was "probably one of the most prominent representatives" to adopt the Neo-Confucian "rationalistic" doctrine of "expanding knowledge through the investigation of things" to study nature and herbs (1986: 181). The great Neo-Confucian thinker Zhu Xi served as a model for Li Shi-Zhen. The title and structure of the *Bencao Gangmu* clearly refer to Zhu Xi's *Tongjian Gangmu*. Moreover, Li conscientiously followed Zhu Xi's maxim that one has to investigate things on one's own in order to arrive at a new and clear understanding of them (Unschuld, 1986, pp. 162, 181). The *Bencao Gangmu* was not at all limited solely as a work on materia medica, but rather a great encyclopedia including the knowledge of areas from pharmacology to general medicine, and from science to culture, in which many recordings were first gathered by Li himself.

The most significant reason for considering Li Shi-Zhen a Confucian physician and medical scholar lies in the fact that he analyzed the practice of human drugs from a Confucian ethical view of point. He applied the key Confucian moral concepts such as *ren* (humaneness), *yi*[a] (righteousness), *li*[a] (reason), and *jun zi* (gentleman) to criticize some of the uses of human drugs. In his general introduction to the section on human drugs, Li remarked:

> In the *Shengnong Bencao* there existed only one medicinal from human beings, that is, hair. The reason for this is to distinguish human beings from any other materials. Unfortunately, the magicians of the later generations even engaged human bones, flesh, gall and blood as medicine. This really contradicts with *ren* (humanness). In this section, I will not omit anything that has been used by someone. However, I will relate in detail only those that do no harm *yi*[a] (righteousness) and leave those miserable, ruthless, wicked and dirty to be described just sketchily. Nevertheless, for the latter I still record them under every entity (Li, Vol. 52: General Introduction).

Apparently in face of human drugs Li Shi-Zhen found himself in a dilemma. As a historian of nature and a physician, he considered it his obligation to gather and include as much knowledge and information as possible on the subject no matter what his own moral belief was. But, as a Confucian, he could not write about what was contradictory with the Confucian moral doctrines. According to Confucius, neither can one talk about even nor listen or see what is contrary to *ren* and *li*[b] (ritual). As a consequence of compromise, Li recorded all the names of human drugs that had been used or at least written down by someone, but dealt with them with careful discrimination. Li did not include any prescription for twelve of the thirty-five human drugs: fetal urine of the infant, dirt from the kneecap, urinary calculi, gallstone, perspiration, tears, breath, the yin soul of a hanged person, facial hair, penis, flesh, and the "mummy." For tear and sweat, Li did not record any therapeutical effects. Li doubted the validity of the historical description in the ancient text on the "Chinese mummy." He attached it as the final human drug "in order for the erudite person to make correction" in future (Li, Vol. 52: Mummy).

For Li Shi-Zhen, it was extremely important to distinguish human bones, flesh, blood, and gall from hair, nails, urine, etc. He condemned the medical use of the former in most situations. In the ancient medical texts it was said that the people in the northern tribes of China apply human bile to the wound in battle. Li commented: "It is the emergent method to save the wound only in the killing field" (Li, Vol. 52: Human Gall). This means that one should avoid using this in other situations. Nevertheless, Li considered collecting and drying human gall for medical use acceptable. The principle here is that "[i]t should not harm *li*[a] (reason)" (Ibid.). Li Shi-Zhen was furious about inhumane action of military people's using human galls. "Some cruel warriors go so far as to take the galls and drink them with liquid as soon as they killed people. They claimed that this could make one brave. This is a ridiculous technique in army. The gentleman should never do it" (Ibid.).

From his anti-Daoism attitude, Li Shi-Zhen reproached the alchemical use of the first menstrual blood of virgins as discussed above. Moreover, from his Confucian moral standpoint he found it unacceptable to use blood in general in medical prescriptions.

> The first who invented this prescription was the most inhumane ("*shen bu ren yi*[a]"). Some even drank human blood with wine. These are the people the Heaven will kill (persecute) and they will definitely get retribution one day. Thus it is not necessary for us to censure them. Among all the prescriptions that used human blood, I will record only those not contrary to *li*[a] (reason, law) in my work (Li, Vol. 52: Human Blood).

In China, medicine is usually called *renshu* (the art of humaneness or humanity). Li did not think that medically using the human bone, even that from the dead, was ethically acceptable. First, it is an age-old custom and virtue to bury human bodies including bones. Second, even animals do not eat their own bones. Third, the bone of even the dead has consciousness.

> The ancients considered it *rende* (humane and virtuous) to cover exposed bones. In doing so one always obtained supernatural reward. But the magicians were concerned only with profits and even gathered human bones to make medicine. How can *renshu* (the art of humaneness) do things like this? Dogs do not eat the bones of dogs, how can a human being eat human bones? (Li, Vol.52: Human Bone)

Based on the idea of filial piety, Li continued, "If the father's bones are exposed, they will soak into the ground as soon as his natural son cuts himself and sprinkles his blood on them" (Ibid.). He further quoted a story from the *Yuyang Zhazu* (The Yuyang Miscellany) to prove that the bone, even separately from the living body, had consciousness.

> A man in Jing-Zhou broke his shinbone. Zhang Qi-Zheng gave him a medicinal wine to drink (as an anesthetic), pierced his flesh, removed a chip of bone, and applied a salve. The patient recovered, but three years later his leg was aching again. Zhang said, "That is because that bone I took out is cold." They looked for it, and it was still under the bed. They washed it in an infusion, wrapped it in silk floss, and put it away, whereupon the pain stopped." Such is the responsiveness of *qi* (the energy). Who says that dry bones have no consciousness? "The benevolent" (that is, doctor) should be aware of this (Li, Vol. 52, Human Bone).

The empirical "fact" and medical theories were often intertwined into the ethical consideration to limit the use of human drugs. By quoting an ancient author Chen Cheng, Li Shi-Zhen tried to show that the medical use of bregma was not effective clinically.

> There was only one human drug, i.e., hair, in the *Shengnong Benjing*. All the other human drugs originated in the physicians or magicians of the later generations. Their ideas are absurd. Recently I saw doctors who used the bregma to treat "cadaver vector disease." But no any effectiveness has been witnessed. Cruel and hurting the god! Truly, not the intention of the humane person. A humane physician must do his best to avoid using it, that is, to use other drugs instead of this. If one has to use, he should choose those that are buried for many years and free from the *qi* of cadaver. (Li, Vol. 52: Bregma)

Here it is very difficult, if possible at all, to distinguish the moral judgment from the factual statement. The empirical fact of the drug's ineffectiveness seems to be subordinate to Li's ethical opposition.

Actually, it is almost impossible to judge the empirical efficacy of the drug with accuracy since "cadaver vector disease," which is believed to be caused by the "ghostly *qi*," is a very ambiguous entity of illness.

On the one hand, Li Shi-Zhen took quite a critical position with regard to human drugs. On the other hand, he was rather ambivalent toward some of them. His views on the medical uses of human flesh and placenta well illustrate these two aspects. The great Confucian scholar-physician strongly opposed the former but felt deeply uncertain to the later.

In the history of Chinese medicine, Li Shi-Zhen was actually the spokesperson against using human flesh as medicine. He especially and explicitly condemned the medical use of human flesh. He said:

> [Even though it is Cheng Chang-Qi who first recorded the therapeutic effects of human flesh,] there already existed the practice of cutting the thigh or the liver before Cheng. People put blame on Cheng for the malpractice because he wrote it down in the book without expounding the ideas to do away with the delusion. Can one speak of materia medica lightly? Alas, one's body, hair and skin come from one's parents and one should not damage them. Even if the parents are suffering from the serious illness, how can they be willing to allow their offsprings to impair their bodies and limbs? How can they have their own bone and flesh? This [taking human flesh as medicine] is the belief of the fool (Li, Vol. 52: Human Flesh)

From this citation, it is very clear that, for Li, using human flesh as medicine was definitely unacceptable. Taking the flesh of the offspring is equal to eating one's own bone and flesh since the body of the child or grandchild is linked with the parents as flesh and blood. Using one's own flesh as medicine or as the material that makes the medical effects of other drugs possible for the purpose of healing one's parents is not morally permissible either. To Li, maintaining the integrity of one's body seems to be a more fundamental moral requirement of filial piety than cutting one's flesh to cure the parent. Here, Li has raised a very important ethical question: Is it ethically justifiable if someone damages one's body for the sake of the parent's well-being? As stated in *Xiao Jing* (The Classic of Filial Piety), it is actually a fundamental moral requirement of filial piety that one should not damage one's body, hair or skin since these come from one's parents. Li attempted to overrule the practice of cutting one's flesh to heal the parent by attributing to the same moral ideal or principle – filial piety. However, Li did not give further exploration to the ethical dilemma, nor did other ancient physicians.

By quoting an contemporary author, Li Shi-Zhen continued to criticize the medical employment of human flesh. It is very clear that according to Confucianism human sacrifice for the purpose of healing is definitely morally and legally wrong.

> The mother of Jiang Bei Er was seriously sick. Jiang cut flesh from his ribs and had his mother take it. But the disease was not healed. He then prayed to god and promised that he would be willing to kill his own son to thank god. His mother recovered from the disease. Jiang really killed his three-year-old son. When hearing about the affair, the Tai Zu Emperor was angry because the activity of Jiang repudiated the normal human relationship (*lun*) and destroyed the natural law (*li*[a]). [The punishment for Jiang's action is] to be flogged with a stick and exiled. The emperor ordered [the officials at] the Department of Ritual to discuss the issue. [They concluded:] "The offspring who really attend and give presents to the parents should request a qualified physician if their parent is ill. It is understandable that, due to the most sincere and earnest feeling, one cannot help calling on Heaven and praying to god. If someone lies on ice and cuts the upper leg, those are foolish persons. ... Some did the world-and-custom-stirring thing with the hope of obtaining praise or escaping the corvee. Someone even killed the child. Nothing runs more counter to the *Dao* and harmful to life than this. From now on, this sort of action [like cutting the upper leg to treat the parent] will no longer be cited in dispatches" (Li, Vol. 52: Human Flesh).

Li Shi-Zhen extolled: "Alas, the doctrine established by the saint is universally applicable through all the ages. It should be this way, [i.e., to prohibit the behavior of cutting flesh or even killing the child]" (Ibid.). In ancient literature it was recorded that some soldiers would eat human flesh in battle. They called the human being "*liangjiao yang*" (the sheep with two feet) and human flesh "*xiang rou*" (thinking meat). Li sternly refuted: "These people are bandits who have no humanity. It is not worth punishing or putting them to death" (Ibid.).

It is interesting to note that, in his opposition to the medical use of human flesh, Li Shi-Zhen followed and repeated the official standpoint of his times. It is more interesting to note that the real cause for the emperor to reject and even prohibit the behavior as an expression of morality, i.e., filial piety, seems to lie not in the moral consideration of Confucianism, but in the fact that many people did this in order to avoid corvee and military service.

With regard to the medical use of placenta, Li Shi-Zhen was much more ambivalent or uncertain than he was with human flesh. He did not explicitly argue that it is ethically problematic to employ placenta as medicine. But he considered it to be a barbarian custom and sighed with a kind of regret that there are few not like barbarians. After citing an ancient author on the importance of carefully burying the placenta for the future of the baby, Li agreed that this theory accords with the law of nature.

> In the *Zhuanyou Lu* (Recordings of Exhausted Journeys), Zhang Shi-Zheng remarked: "The long barbarians in Ba Gui cook the placenta with five flavors and eat together when the boy baby is born. This originated in imitating animals which eat the placenta when the baby animal has been born. But the animal is not a human being." In the *Xiaoer Fang* (The Prescriptions for Children) Cui Ying-Gong remarked: "The placenta should be buried in auspicious places according to the *de* (virtue) of the Heaven and the Moon. The body will attain a longevity if its placenta is buried deeply and covered tightly. The child will suffer from madness and withdrawal if its placenta is eaten by the pig and dog; sore and lichen if eaten by insects and ants; fearing death if eaten by birds; putrefying clove sore if thrown into fire. It is also prohibited to put it near the temple, dirty water, and the well." ... This is the *li*[a] (nature, reason, law) of the nature. But now the placenta is steamed, boiled, proceeded [such as soaked, dried, calcined, baked, carbonized, roasted], blended with drugs, and pounded with pastry. Although the placenta can supplement the human being due to their being of a kind, does the human eating the human alone not violate the prohibitions Mr. Cui set up? How few the people who can be distinguished from the barbarians in Liu Qiu! (Li, Vol. 52: Human Placenta)

As a matter of fact, it was a very old custom to bury the placenta carefully in China. It was believed that the date, the place, and the direction of burying determined the child's fortune. The custom still remains in some places of southern China.

Yet, Li Shi-Zhen quoted in length what Wu Qiu, the inventor of the Great Creation Bolus, said about the magic effects of human placenta without any ethical criticism (see the citation in last section). In addition to those medical results promoted by Wu, Li agreed with other ancient physicians that human placenta, combined with the different drugs, could also treat many other diseases, including the long-term insanity, "five taxations and seven damages," epilepsy, *gu* toxin (poisons of a legendary venomous insect), red eyes and eye screen (Ibid.).

There is other evidence to indicate Li Shi-Zhen's ethical ambivalence regarding human drugs. He recorded thirty-five human drugs, but it is difficult to judge whether he did so as a medical scholar and historian of nature whose top moral obligation is knowledge or whether he thought the human drug ethically justifiable. For instance, Li reproached the medical use of blood. At the same time, he included some prescriptions of human blood and affirmed its therapeutic effects for such ailments as prolonged hematemesis (vomiting of blood), prolonged epistaxis (nosebleed), internal hemorrhage from wounds, postpartum nausea with abnormal blood loss, scarlet birthmark in infants, warts in infants. Li strongly rebuked using human gallbladders as medicine. Yet, he believed that the gallbladder could treat long-term malaria, food choking and wounds from mental weapons. While Li clearly stated that "the human penis is not a drug" (Li, Vol. 52: Penis), he still quoted an anecdote from the ancient literature in which a sort of therapeutic affect had been confirmed.[7]

In summary, the Confucian physician-scholar Li Shi-Zhen's attitude toward human drugs has two characteristics. First, unlike Daoists, Li was very sensitive to the ethical aspects of human drugs. In fact, he was the first Chinese physician who explicitly discussed ethical issues of human drugs according to the Confucian moral ideas like *ren* (humaneness), *yi*[a] (righteousness), *li*[a] (ritual), and *li*[b] (reason or natural law). For Li Shi-Zhen, some human drugs, like blood, gallbladder, bone, and flesh are much more ethically problematic than the others, like hair, nails, urine and urinary sediments. Obviously, this distinction is extremely significant for the ethical analysis of human drugs. Second, even though Li considered some human drugs ethically unacceptable, more often he was ambivalent. He was not very clear or consistent about the ethical issues of human drugs. He did not establish a systematic Confucian version on human drugs nor any absolute limitation against the practice from the standpoint of Confucianism.

V. FROM HISTORY TO THE PRESENT

Up to this point, I hope that I have shown three things. First, an interpretative approach requires that one should not just ignore the apparent absurdities in the knowledge and practice of human drugs. Secondly, a lot of complex historical and cultural factors are involved in the invention and popularity of human drugs. Thirdly, from the Confucian ethical perspective, the great sixteenth-century physician-scholar Li Shi-Zhen took a cautious and critical attitude toward the medical use of some human drugs such as flesh, bone and blood. In this final section, I will discuss some theoretical and practical implications of a historical-ethical analysis of ancient Chinese human drug practice for today.

1. The Importance of Historical-Cultural Context in Addressing Medical Ethical Issues

In section II, I listed emphasizing the historical-cultural context as one of the characteristics of an interpretative approach to medicine and medical ethics. In section III, I showed some aspects of the historical-cultural context of Chinese human drug practice. In particular, I discussed the role of the Chinese virtue *xiao* (filial piety) in the practice of cutting one's flesh to heal the parent by mixing it with other medicine and the role of the Chinese illness *xulao* (general weakness and exhaustion) in using human placenta. I must emphasize that I do not mean to suggest that the conception of filial piety would necessarily result in the practice of employing human flesh as medicine. As in the case of human flesh, my point of introducing the cultural background of the medical use of placenta is not to suggest that the Chinese medical and social understanding of *xulao* would necessarily result in the employment of placenta, nor to suggest that the use of placenta is justifiable because general weakness and exhaustion is a actual disease in Chinese culture. Rather, what I try to assert is that, in our discussion about the ethical issues of human drugs in particular and medical practice in general, we should take the historical-cultural context into account.

Given the great importance of filial piety in traditional Chinese moral and social life, it should become at least understandable why the action of cutting one's flesh to treat one's parent was quite popular and cited in so many official documents of imperial China. We can surely say that the Chinese practice of employing human flesh for therapeutic purposes is not a moral evil, although maybe not a great moral deed either. If we were to take away the historical and cultural context from the practice of human drugs, the medical use of human flesh in China would appear to be barbaric, primitive, and possibly even cannibalistic.

It is almost commonplace nowadays to say that medicine, like the Chinese practice of human drugs, is inevitably contaminated with the historical and cultural factors. But the mainstream discourse of contemporary bioethics seems to be indifferent and sometimes even antipathetic to the historical and cultural context of ethical issues in medicine and health care. A series of myths or basic assumptions in contemporary bioethics, such as the myth of the autonomous self, the myth of the context-free and universal principles, and the myth of coherence, need to be more critically examined. Three turns, which can be respectively called historical, anthropological, and interpretive, taken together, constitute a necessary and effective antidote to these modern ethical myths. A historical turn in bioethics requests tolerance and appreciation for "imprecision and irregularity" in medical morality by "welcoming into the discussion of the moral dimensions of medicine the often obscure traditions of medicine's historical and mythic past" (Jonsen, 1990, p. ix). An anthropological turn aims to bring to bioethics a "critical approach to the cultural roots of ethical systems and to the cultural process of moral action." And it encourages empathy for the diversity of moral meanings and for "contradiction, contestation, and paradox in a world without an intrinsic moral rule" (Kleinman, 1995, pp. 57, 60). An interpretive turn in bioethics seeks to engage the commonplace, the plausible, the ambiguous, and the uncertain, as well as personal, responsive aspects of human moral life related to healing and curing. The way of interpretation as discernment in medical ethics acts as "a first cousin to practical reason and casuistry" and "a sibling of hermeneutics and thick description" (Carson, 1990, p. 56). When these three approaches are brought together, we are able to face squarely the historical, cultural, and practical dimensions of medical ethical issues. We are then able to address more accurately such questions as the following: how moral agents and their activities are shaped by the historical, cultural, and social context? How do medical moralities work in reality? How can ethical ideals and principles be actualized by the historically and culturally bound individuals in historically and cultural bound circumstances.

In light of the indifference and even antipathy of contemporary mainstream bioethics to the significance of historical and cultural factors in medical ethical issues, the need to integrate anthropological, hermeneutical, and historical perspectives into the scholarship of medical ethics cannot be overemphasized. These perspectives can help us uncover complexities of medical practices, suspend our judgments to the difference, view our own practices in a new light. Nevertheless, as I have pointed out elsewhere (Nie, 1998), paying attention to the importance of cultural differences never means justifying everything in a particular society. Equally, doing one's best to appreciate the historical contexts of medical theories and practice does not imply a justification for that which is understood in history. In the face of the potential dangers of cultural relativism and ethical nihilism, this point also cannot be overstated. Anyway, if ethical exploration is to have implications on reality, this exploration must start from reality – its historical-cultural context. Otherwise, the ethical analysis will hardly be relevant to the everyday practice.

2. From Human Drugs to Organ Transplantation

Beyond doubt, there are striking differences between the Chinese uses of human drugs and the contemporary high technology of organ transplantation in biomedicine. For instance, most human drugs were believed to be able to cure various diseases, but rarely supposed to have the effect of saving life while tissue and organ transplantation is usually thought to be able to save lives. In the practice of human drugs, body parts, excreta, and appendages are taken in or applied topically like other medicinal plants and minerals, while in transplantation the diseased organ or tissue is replaced by an alive and healthy one. As one of the most dramatic high medical technologies, organ and tissue transplantation has other remarkable features. Its success depends not only on many physicians and other medical specialists working together as a team but also on long-term medical intervention to deal with the post-operative problems. Like all advanced biomedical procedures, its cost is very high.

However, no matter how new and advanced the technology of organ and tissue transplantation is, it shares a fundamental character with the old and strange-sounding practice of human drugs; that is to say, both involve with the therapeutical use of human body parts as mentioned at the beginning of this paper. The striking differences between human drugs and organ transplantation do change this crucial commonality: human body parts being directly used for the purpose (or in the name) of healing and curing. For ethical discussion, this shared feature is especially significant, even more so than their differences are.

It is my assertion that, to a great extent, the use of human drugs is the ancient counterpart of organ and tissue transplantation; or vice verse, the advanced medical technology can be seen as a kind of modern form of the practice of human drugs. If this is the case, then the historical exploration of the Chinese practice of human drugs is not limited to human drugs *per se,* nor has significance merely for the history of medicine. The historical-ethical discussion of human drugs has the direct practical implications, especially to China. This is not only because some human drugs such as placenta and hair are still a part of contemporary Chinese pharmacology, but also because China has started to develop organ transplantation in recent years. Without proper understanding of the traditional practice and the ancient physicians' moral concerns about human drugs, it would be impossible to achieve a clear concept or picture of contemporary Chinese attitudes toward and views of organ transplantation.

Moreover, the ethical correlation between human drugs and the advanced medical technology of organ transplantation calls for a more explicit articulation and a more systematic framework for the medical use of human biological materials drawn from the Confucian traditions. This should become an urgent agenda for scholars in the field of Chinese medical ethics in China as well as in the West. In section IV, I presented the Confucian physician Li Shi-Zhen's critical position towards human drugs. However, neither the author of the *Bencao Gangmu* nor any other medical scholar in traditional China offered a fully developed ethical perspective on human drugs. Li did not criticize the use of some human drugs so much from a coherent Confucian ethical theory as from moral instincts or common sense arising in Confucianism and Chinese culture. In the last part of this section I will give a little bit further discussion on how to develop a Confucian framework of the medical use of human body parts – from human drugs to organ transplantation.

3. Are Human Drugs Ethically Justifiable?

Is the use of human drugs ethically justifiable? The medical uses of human flesh and placenta manifest the most puzzling ethical issues in the Chinese human drug practice. If it is shocking that the use of human flesh as medicine had been popular in some periods of pre-modern China, then it is troubling that the human placenta is still widely employed as an important tonic medicine in present China. In the intellectual and traditional-style medical circles of contemporary China, following Li Shi-Zhen, people consider the practice of cutting one's flesh to heal one's parent not only morally wrong but also an illustration of the backwardness and ignorance of Confucianism. However, quite surprisingly, few physicians and scholars have raised the question whether the medical use of human placenta in particular and the practice of human drugs in general is ethically acceptable. But why is the cutting of one's flesh to heal one's parent wrong while the medical use of placenta and other human drugs like hair and nails is not ethically problematic? Is the therapeutical use of human flesh really nothing more than backward and ignorant?

Actually, the question whether the medical use of human drugs is ethically justifiable includes two different, though related, issues. First, is the therapeutic use of some parts and appendages of human body is morally acceptable in general? Secondly, is the particular procedure of employing a human drug ethically justifiable?

Suppose that a woman is suffering from a malignant tumor. Some evidences have led her son to believe that if he cuts a piece of flesh from his upper arm or thigh to mix with other medicine, his mother will probably recover from her serious disease. Finally, the son voluntarily cuts his flesh and the woman, knowing the nature of her medicine, takes the herbs which have been boiled with her son's flesh. According to the standard view of contemporary Chinese intellectuals and doctors, the action of the son is no more or less a expression of his "stupidness and benightedness" and "foolish filial piety." However, I disagree with this point of view. Ethically speaking, there is no fundamental ethical difference between cutting one flesh to cure his mother and donating one of his kidneys to save his mother's life. One may argue that the former is not as empirically effective as the latter. But this argument does not show that the former is ethically wrong. It seems to me that, if the medical effects of human flesh and placenta are valid as claimed, then the use of these two human drugs is morally justifiable; more exactly speaking, permissible. That is, there are no legitimate moral reasons to prove that the medical use of human drugs is ethically unacceptable. The use of most human drugs in Chinese medicine is not intrinsically evil or objectionable. In other word, like organ transplantation, human drugs in general are morally acceptable or permissible, if not required.

Nevertheless, I have strong reservations against supporting the medical use of human flesh and placenta and many other human drugs. The reason is not that the claimed medical effects of the most human drugs are problematic, if not completely unfounded. For Li Shi-Zhen, the employment of the some human drugs is problematic because of ethical reasons, especially in the cases of human flesh, blood, and bone. But he believed in the clinical effects of most of them, especially in the cases of placenta, hair, and others. Li never expressed any doubt about the medical effects of the Great Creation Bolus. Actually, he spoke highly of the formula's composition and pointed out that the bolus, as a tonic recipe, "can be taken even without placenta" (Li, Vol. 52: Human Placenta). On the contrary, I do not oppose the medical use of human drugs in general because it is morally unacceptable; it is not. Rather, I am not convinced that they are clinically effective. Nevertheless, with regard to the special nature of human body, I find Li Shi-Zhen's cautious, ambivalent, and critical attitude to human drugs has special theoretical and practical significance even for today.

Let me say a few more words about my personal feeling of human drugs. As Li Shi-Zhen, I feel disgusted about most human drugs, with only a few exceptions like hair and nails. I am positive that if I suffer from a serious chronic disease, I will not allow my daughter to cut flesh from her arm in order to cure me. I am also positive that if I feel weak and deficient, I will not allow my parents or my wife to get and cook a human placenta for me to take. However, I am not as sure that when one or both of my parents has suffered from serious chronic diseases for years, I will refuse, without second thought, to cut my own flesh to mix with other medicine for the purpose of curing my parents if the traditional-style physician and my uncle suggest this and if many people believe in the effectiveness. Nor am I as sure that, I will never agree to use human placenta if my daughter is suffering from weakness, learning disability, and epilepsy and if her Chinese physician prescribes her the Bolus of Great Creation, and if my wife believes in the efficacy of placenta.

I have pointed out above that the medical use of flesh and placenta in particular and human drugs in general is not morally unjustifiable. But this does not mean that the traditional practice of human drugs is never morally problematic, just as the justification of organ transplantation does not imply that harvesting the organs by force is morally acceptable. In fact, the Chinese human drug practice has a dark side. One aspect of the dark side is the Daoist practice of inducing or forcing virgin girls to have sex and obtaining milk from unmarried young girls, to which Li Shi-Zhen harshly criticized from a Confucian perspective and common Chinese moral sense. By modern standard, this practice is child abuse and sexual assault.

But what is ethically problematic and wrong is often not so much about the use of human drugs as about the procedures of employing these drugs. From the perspective of contemporary bioethics, an important ethical transgression in the practice of human drugs is the violation of informed consent. In using human drugs, it was sometimes medically required to prevent the relevant agents from being aware of the medical procedure; otherwise, the medicine would not have effects. For example, it was said that although human placenta could cure the general weakness and exhaustion of women, the doctor warned: "Do not let the woman know." Here, from the context of the word, "the woman" could refer to the one from whom the placenta has been obtained or the one who takes the placenta as medicine or probably both.

The connection of human drugs with organ transplantation calls for a serious study of these dark aspects. In spite of the fact that human beings never really learn the lessons of history, the historical study can at least help people to become aware that we often repeat history. It is true that a historical study of the traditional Chinese human drugs practice can never entirely prevent the unethical things from occurring in the contemporary Chinese practice of human drugs as well as organ transplantation. Yet, the study requires us to ask the following question: Is China repeating the same kind of ethical mistakes that have occurred in the ancient practice of human drugs when contemporary Chinese physicians are allowed or pushed to harvest organs from would-be-executed prisoners?

4. Toward a Confucian Framework for the Medical Use of Human Biological Materials

Due to the lack of research and exploration, it is far from possible here to give a detailed description of a Confucian perspective on the medical use of human biological materials including human drugs and organ transplantation. Yet, before ending this paper, I would like to give a rough sketch on what this Confucian framework for the medical use of human biological materials would look like.

In order to discover and develop a Confucian ethical theory of the medical use of the human body, we need first of all to be more humble with regard to the Confucian thought traditions. This humble and conversational demeanor toward history and tradition is what contemporary philosophical hermeneutics suggests and Confucianism itself encourages. In this century, Confucius and the doctrine he initiated have been treated either as not relevant or as negatively relevant to the present time. They are rarely treated as the inspiring partners of intellectual dialogues. Even though Confucianism was the orthodox political-philosophical-religious system in pre-modern China, modern Chinese intellectuals have harshly rejected it as the typical old, backward and "feudal" obstacle to science, democracy, the Enlightenment, modernity, and modernization. The popular slogan "Down the Confucian Shop" was not just the clarion call in the era of the May Fourth Movement, but a *leitmotiv* in the unharmonious symphony of twentieth-century Chinese culture. To many Westerners, Confucius and Confucianism are not only archaic but also foreign and parochial. But if we – modern Chinese and Westerners – are humble enough to listen carefully and imaginative enough to interpret creatively, then the ideas and teachings of Confucius and his followers will shed much light on the problems and issues we are face today, including those in medicine and health care. I hope that my essay, along with this entire volume, have demonstrated that Confucianism is relevant to contemporary medical ethics.

The human body never means the same thing to people in different cultural traditions and historical ages. Consciously and unconsciously, any ethical analysis on the medical uses of human biological materials – whether human drugs or organ transplantation – is based on a certain understanding of the body. The Confucian and Chinese medical images and metaphors of the human body provide the foundation for a Confucian perspective on the subject. Confucianism has greatly influenced the Chinese medical perception of the body, and actually constitutes an organic part of the philosophy in which Chinese medicine evolved. As a result, to a great extent Confucianism and Chinese medicine have a common discourse and share some fundamental assumptions on the structure and functions of the body, its relations with society and nature. Although it is problematic to claim that these ways of thinking are totally alien to the West, Confucianism and Chinese medicine do perceive the body in a way rather different from that of orthodox modern Western philosophy and medicine. For instance, in contrast with the popular Western view in which the body is understood as a fixed material, in the Chinese perception the body is never a static existence, but a form of *qi* (vital energy), always in flux. In the physiological as well as the religious, spiritual, moral, and social senses, the body has also been viewed by the Confucians and Chinese doctors as "the net in a web without the weaver," "a microcosm closely corresponding with the macrocosm," "the holy vessel," "a vehicle or conduit."

Since the human body is of a special dignity and even sacredness according to Confucianism, any intervention in and manipulation of it must be carefully examined. Confucians hold and promote numerous virtues for achieving personal perfection and self-cultivation. Most personal virtues in the Confucian moral system are also moral principles and thus they can be extended to analyze social practices. In Confucianism, *ren* (humanity, humaneness, benevolence, the basis of all goodness), *yi*[a] (righteousness, just), *li*[a] (ritual, propriety), and *zhi*[b] (wisdom, knowledge) are "the Four Beginnings" of a full human being. I think that a Confucian framework on the medical use of the human body and its parts can be established on these "Four Primordial Virtues." These four Confucian ethical ideals are able to help make inquires and give insights into the following issues: What is the motivation for and what are the consequences of the medical use of human biological materials? Does this use accord with the principle of *ren* and satisfy the moral requirements of medicine as the art of humanity? Is it empirically correct or wise to use the human body or its parts on the particular conditions? If the use is humane and wise, are the adopted procedures righteous and just? If the use is humane, wise, and just, what kind of particular ritual should be developed to make the human activity fully "human"?

Center for Bioethics
University of Minnesota
Minneapolis, USA

NOTES

* I thank Faith Lagay and Ruiping Fan for their valuable comments and generous help with this essay. Ronald Carson's commentary is greatly appreciated.

[1] All the English translations of Chinese sources in this paper are mine. There is no complete English version of the *Bencao Gangmu* yet. In rendering the materials from *Bencao Gangmu*, I have referred to and benefit from some contemporary works, especially Cooper and Sivin (1973).

Since there are many different printings of the *Bencao Gangmu*, I will indicate the source by putting the number of the volume and the name of the drug in parentheses, instead of the page number.

Li Shi-Zhen did not actually write the whole work of *Bencao Gangmu*, but compiled it. He systematically collected all materials available to him on materia medica and related fields. In the entry on pubic hair, besides the knowledge he first gathered, he also used materials from four ancient medical and pharmacological works.

[2] In the *Shengnong Bencao*, all the three-hundred-sixty-five drugs were classified into three classes – the upper, the middle, and the lower. Over two-thirds of them were herbs, trees, fruits, vegetables, and grains.

[3] The title of this book has been translated differently as *The Great Pharmacopoeias*, *The Systematic Pharmacopoeias*, *Compendium of Materia Medica*, *Materia Medica Arranged According to Drug Descriptions and Technical Aspects*, and *The Pandects of Pharmaceutical Natural History*.

[4] There are three theories about the role of the gender. First, a female patient should use the placenta of a boy baby and while a male patient should use that of a girl baby. Second, a female patient should use the placenta of a girl baby and the male patient that of a boy baby. Third, sex does not matter in employing afterbirth as medicine (Li, Vol. 52: Human Placenta).

[5] The story tells of a king who suffered from a lai sore in the whole body. His physician said that the medicine would not be effective unless blended with flesh from his own children. Among his three daughters, only the youngest princess Miao Yin was willing to cut flesh from her arm to cure her father's disease. Later, the king had the problem with his eyes. Miao Yin cut one of her own eyes and her father's eye disease was cured again (Cited in Ma, 1993, p. 446).

[6] With regard to the influence on medicine and science, there are some differences between classical Confucianism and Neo-Confucianism. Generally speaking, both Confucianism and Neo-Confucianism were basically humanistic-based, rather than scientific-based, and this-worldly oriented, rather than other-worldly oriented even though the humanities and sciences, this world and other world tendencies did not necessary excluded from each other. Nevertheless, according to Needham and Wang (1956), classic Confucianism and Neo-Confucianism took somehow different attitudes towards science and hereby exerted different influences on the historical development of Chinese science and medicine. The main and almost exclusive concentration of interest of classic Confucianism on human affairs and relationship in social life might be unfavorable for the scientific inquiries into non-human or natural phenomena although its rationalistic tendency might be advantageous to science at the same time. The *Lun Yu* (*The Analects*) (12: 20) told that Confucius never spoke of the following four subjects: extraordinary things (natural prodigies), unnatural forces, disorders (in nature), and spiritual beings. In contrast, the Master frequently discoursed the Odes, the history and the maintenance of the Rites, culture (letters), and the conduct of affairs. Yet, empirical sciences would never have arisen and developed without careful study of these phenomena about which Confucius seldom spoke. For Needham and Wang, the Neo-Confucianism that culminated in the great twelfth-century Chinese philosopher Zhu Xi was

"essentially scientific in quality." Its system of thought and method of inquiry were more advantageous to scientific investigation and technological inventions and actually "accompanied by a hitherto unparalleled flowering of all kinds of activities in the pure and applied sciences themselves" (1956, p. 496). More significantly, Chinese organic naturalism, based originally on a system of correlative thinking initiated by the Daoist philosophers and systemized in the Neo-Confucian thinkers, provides the theoretical foundation for a new world-view different from the mechanical one in modern Western sciences (Needham and Wang, 1956, p. 505).

[7] The story states: "A man cut his penis because of the revelation of adultery. But the injury had been bleeding for months. Someone else asked the man to get his cut penis, pound it into powder, and drink with wine. Only a few days later, his wound was cured." (Li, Vol. 52: Human Penis)

REFERENCES

Primary Source:

Li, S.Z.: 1988, 1592, *Bencao Gangmu*, Chinese Bookstore, Beijing, China.

Secondary Sources:

Carson, R.A.: 1990, 'Interpretive bioethics: The way of discernment,' *Theoretical Medicine* 11:51-59.

Carson, R.A.: 1995, 'Interpretation,' in Warren T. Reich (ed.), *Encyclopedia of Bioethics*, Revised Edition, Simon & Schuster Macmillan, New York. pp.1281-87.

Carson, R.A: ms., *Medical Ethics as Moral Works.*

Cooper, W.C. and Sivin, N.: 1973, 'Man as a medicine: Pharmacological and ritual aspect of traditional therapy using drugs derived from the human body,' in Shigeru Nakayama and Nathan Sivin (eds.), *Chinese Science: Explorations of an Ancient Tradition*, The MIT Press, Cambridge, Massachusetts.

Dreyfus, H.L. and Rabinow, P.: 1983, *Michael Foucault: Beyond Structuralism and Hermeneutics*, The University of Chicago Press, Chicago.

Foucault, M.: 1973, *The Birth of the Clinic: An Archaeology of Medical Perception*, trans. by A.M.S. Smith, Vintage/Random House, New York.

Geertz, C.: 1973, *The Interpretation of Culture: Selected Essays.* Basic Books, New York.

Gulik, Robert H. van: 1961, *Sexual Life in Ancient China*, E.J. Brill, Leiden, Netherlands.

He, Y.M., Nie, J.B., and Wang, X.D., *et. al.*: 1990, *Chayi, Kunhuo Yu Xuanzhe: Zhongxi Yixue Bigiao Yangjiu* (Differences, Perplexities, and Choices: Comparative Studies on Chinese and Western Medicine). Shenyang Press, Shenyang, China.

Jonsen, A.R.: 1990, *The New Medicine and the Old Ethics*, Harvard University Press, Cambridge, Massachusetts.

Keys, J.D.: 1976, *Chinese Herbs: Their Botany, Chemistry, and Pharmacodynamics*, Charles E. Tuttle Company, Japan.

Kleinman, A.: 1980, *Patients and Healers in the Context of Culture: An Exploration of the Borderland between Anthropology, Medicine, and Psychiatry.* University of California Press, Berkeley, California.

Kleinman, A.: 1988, *The Illness Narratives: Suffering, Healing, and the Human Condition.* Basic Books, New York.

Kleinman, A.: 1995, *Writing at the Margin: Discourse between Anthropology and Medicine*, University of California Press, Berkeley, California.

Kuhn, T.S.: 1970, *The Structure of Scientific Revolution*, 2nd Ed., The University of Chicago Press, Chicago.

Kuhn, T.S.: 1977, *The Essential Tension: Selected Studies in Scientific Tradition and Change.* University of Chicago Press, Chicago.

Ma, B.Y.: 1993, *Zhongguo Yixue Wenhua Shi* (The History of Medicine in Chinese Culture), Shangshai People's Press, Shangshai, China.

Mink, L.O.: 1991, *Historical Understanding.* Cornell University Press, Ithaca.

Needham, J., with Wang, L.: 1956, *Science and Civilization in China*, Vol. II, The Cambridge University Press, Cambridge, England.

Needham, J., with Lu, G.D.: 1983, *Science and Civilization in China*, Vol. V, Part V, The Cambridge University Press, Cambridge, England.

Nie, J.B.: 1989, 'On studying traditional Chinese medicine by modern scientific methods,' *Yixue Yu Zhexue* (Chinese Journal of Medicine and Philosophy) 10(8):22-24.

Nie, J.B.: 1995, 'Scientism and traditional Chinese medicine in the twentieth-century China,'
Yixue Yu Zhexue (Chinese Journal of Medicine and Philosophy) 16(2):62-66.

Nie, J.B.: 1997, 'Refutation of the claim that the ancient Chinese described the circulation of blood: A critique of scientism in the historiography of Chinese medicine,' a paper presented at 1997 annual meeting of the American Association for the History of Medicine, Williamsburg, Virginia.

Nie, J.B.: 1998, 'Coerced abortion in China: Some ethical issues,' *Cambridge Quarterly of Healthcare Ethics* 7(2).

Unschuld, P.U.: 1986, *Medicine in China: A History of Pharmaceutics*, University of California Press, Berkeley, California.

Walzer, M.: 1987. *Interpretation and Social Criticism*, Harvard University Press, Cambridge, Massachusetts.

Xie, Z.F., and Huang, X.K.: 1984, *Dictionary of Traditional Chinese Medicine*, The Commercial Press, Ltd., Hong Kong.

RONALD A. CARSON

INTERPRETING STRANGE PRACTICES

What is one to make of apparently bizarre practices encountered in societies different from one's own? This is the background question raised by Jing-bao Nie in his thought-provoking paper on the use of "human drugs" in traditional Chinese medicine.

Nie enumerates two common responses to this question. According to the "iconoclastic" response, if a traditional prescription has no basis in modern scientific knowledge, it should be dismissed as nonsense. A "progressivist" view, by contrast, acknowledges the superstitious nature of such a prescription, but allows that traditional practices often have their own *raison d'être* and that the challenge is to distinguish the irrational (and therefore dispensable) dross from the rational elements, however odd these latter may seem by current standards. Nie correctly points out that despite the obvious variance between the iconoclastic and progressivist responses, they share the assumption that it is possible to step outside history and tradition to assess the worth and validity of a practice foreign to our experience from the perspective of our (enlightened) way of understanding things. Nie aligns himself with critics of this modernist assumption and offers a post-modernist and interpretive approach to understanding strange practices.

Drawing on the work of Michel Foucault and Thomas Kuhn, Nie points out that what seems nonsensical or superstitious when viewed through a modernist lens may as likely be a function of the lens as of the object of attention. Strangeness may disappear when we come to see a foreign idea or practice as part of a different mind-set or life-world. By reserving judgment, we may come to an appreciation of difference. This is the chief merit of a post-modernist approach. But for Nie, to leave matters at treasuring difference is to stop short of exercising critical judgment. To achieve an appreciation for the sense something makes in its own frame of reference is a necessary step on the way to deciding what sense it makes (or doesn't) at the borders between frames of reference where normative ideas and practices collide or coexist. For this an interpretive approach recommends itself.[1]

Interpretivists acknowledge difference and subscribe to the belief that every view is a view from some particular (unprivileged) place. They also

Ruiping Fan (ed.), Confucian Bioethics, 207–210.

believe that some views are better than others and that through open, persistent, respectful dialogue fresh understanding is possible – understanding that neither coerces consensus nor fudges fundamental differences but rests, however provisionally, on common ground. The interpretive method aims not to discover or demonstrate definitive knowledge but to advance practical understanding. Interpretivists set out from the assumption that we live our lives in webs of meaning, that, in fact, our lives are constituted in some fundamental way by the sense we are able to make of them. This assumption informs my understanding of myself as a late twentieth-century American trying to fathom what ancient Chinese healers might have had in mind in prescribing "human drugs." It also shapes my mental image of what those healers were about in their own time and place, so very different from mine, but not so dissimilar as to be beyond my comprehension. They too, I assume, believed their prescriptions to be meaningful responses to illness. But in what way? How could they have thought that cutting a child's flesh might cure her father's ailment?

Only a literalist of the imagination would look for a cause-effect connection here. Jing-bao Nie ably employs the interpretive method in looking to the cultural context within which human tissue was *materia medica* and asking what social purposes were served by this understanding.[2] That context was, importantly, Confucian. That is to say, certain values and virtues associated with Confucianism were influential in shaping cultural attitudes toward healing practices. As Nie writes, "In the Chinese view, it is not the human drug per se but certain human relationships and virtues that really cure." Filial piety, the animating virtue of the relationship between child and parent, was paramount. So it might make eminent sense, when confronted by a case of parental illness, to recommend that the child do something for the parent. Doing something for one's ailing father or mother might have a healing effect. Undertaking something radical – something self-sacrificial, for example – might well have a radically positive effect. One can readily see how, in a culture in which filial piety was a regnant virtue, cutting one's flesh to cure one's parent could be thought to strengthen bonds between children and parents.[3]

As Jing-bao Nie ably employs the interpretive method, we come to think of the differences between our own healing practices and the traditional Chinese practice of recommending human flesh for medicinal purposes as differences in ranges of meaning available to participants in the two cultures. Such meaning is neither objective (universally valid) nor subjective (culturally relative) but intersubjective. The meaningfulness of a practice resides not only in the minds of the practitioners but also "out there in the practices themselves," practices understood here as "essentially modes of social relation, of mutual action."[4]

Such an interpretive construal of an ancient practice causes us, in turn, to reflect on current practice – in this paper, that of organ transplantation – which as Jing-bao Nie observes, also involves the therapeutic use of human body parts. What moral sense are we to make of the practice of removing vital organs from the body of one person and placing them in the body of another? The interpretivist's answer is: it depends on what those who engage in the practice believe about their bodies, their lives, and their place in nature and history, and on the meaning of the practice to those from different moral traditions who have no enthusiasm for organ transplantation and who must come to terms with those who do, and *vice-versa*. The need to make moral sense of a practice arises only when the "working sense" no longer suffices or when traditions of meaning clash. The interpretivist responds to this need by "bringing incommensurable perspectives on things, dissimilar ways of registering experiences and phrasing lives, into conceptual proximity such that, though our sense of their distinctiveness is not reduced (normally, it is deepened), they seem somehow less enigmatical than they do when they are looked at apart."[5]

For the interpretivist, moral meaning arises from engagements – textual comparisons, ordinary conversations, formal negotiations, heated debates – between "us" and "them." It is in such encounters with difference that one recognizes the merits and limits of "strange" practices, including one's own.

Institute for the Medical Humanities
The University of Texas Medical Branch
Galveston, USA

NOTES

1 For a detailed argument in this regard, see (Carson, 1995).

2 Alasdair MacIntyre has observed that "different institutions, embodying different conceptual schemes, may be illuminatingly seen as serving the same social necessities" (MacIntyre, 1978, p. 299).

[3] Left unconstrained, however, this practice could turn vicious, which may be among the reasons why Li Shi-Zhen condemned the practice and commended the maintenance of bodily integrity as a higher expression of the virtue of filial piety.

[4] See Taylor, 1985, p.36.

[5] See Geertz, 1983, p. 233.

REFERENCES

Carson, R.A.: 1995, 'Interpretation,' *Encyclopedia of Bioethics,* Warren Thomas Rich (ed.), Simon & Schuster, London, Volume III: 1283-1288.

Geertz, C.: 1983, 'Local knowledge: Fact and law in comparative perspective,' *Local Knowledge*, Basic Books, Inc., New York.

MacIntyre, A.: 1978, 'The idea of a social science,' *Against the Self-Images of the Age*, University of Notre Dame Press, Notre Dame, Indiana.

Taylor, C.: 1985, 'Interpretation and sciences of man,' *Philosophy and the Human Sciences*, Cambridge University Press, London.

XUNWU CHEN

A CONFUCIAN REFLECTION ON EXPERIMENTING WITH HUMAN SUBJECTS

Experimenting with human subjects in medicine and biology encounters significant ethical challenges today. There are general questions regarding the justification of the importance of experiments of this kind. For example, is it justifiable for medical or biological experiments to use human persons as *subjects* in general? Are these experiments justifiable in certain specific circumstances (i.e., in a non-therapeutic context)? Is it justifiable in medical or biological experiments to use certain groups of persons (i.e., the infant, the fetus, the unborn, or the prisoner) as subjects? Does an experiment with human subjects impair the dignity and sacrosanctity of the individuals who are *the subjects* of the experiment in particular and insult humanity in general? Can there be any adequate justification for an experiment with human subjects? Is a justification of experiments of this kind only perspectival, historical, and practical? There are also other ethical issues concerning the conditions and procedures of experimenting with human subjects. For example, is informed, uncoerced consent from individuals to become *subjects* of an experiment necessary? Does a differentiated, controlled arrangement of subjects in a medical or biological experiment constitute a kind of discrimination that we normally fight against in real life?

This essay attempts to reflect on these bioethical issues from a Confucian perspective. The term "reflection" is deliberately used here to emphasize that what I am going to present is not a systematic and comprehensive account of Confucian bioethics. Neither the views articulated below exhaust what Confucianism can and will say on the subject matter of experimentation with human subjects nor are these views themselves organized into a systematic theory. It should also be stated at the outset that the Confucianism referred to here is not restricted to views of great Confucianists such as Confucius himself, Mencius, Chu Shi, etc. Rather it includes also what Oskar Weggle calls meta-Confucianism; namely, Confucianism held and practiced in ordinary life by farmers, workers, businessmen and women, scientists, physicians, doctors, etc. (Weggle, 407). In the context of the present study,

Ruiping Fan (ed.), Confucian Bioethics, 211–232.

Confucianism is used in a broader sense. Finally, to avoid vagueness and retain focus, I shall define an "experiment" here as a *controlled, planned* experimental practice in the fields of medicine and biological science; it does not include those inchoate, random, and un-planned practices.

I. THE STORY OF THE *LEGENDARY FARMER* AND THE CONFUCIAN CONCEPTIONS OF MEDICAL AND BIOLOGICAL KNOWLEDGE, HUMANITY, AND SOCIAL PROGRESS

Let me start with the Chinese story about the *legendary farmer*, a historical figure in ancient China. The origin of the traditional Chinese medicine, commonly referred to as "Chinese grass medicine," is normally traced back to the legendary farmer's self-trial of one hundred kinds of grasses (plants). As Shi MaZhen's *The Appendix of the Three Emperors of the Records of the History* tells us: "the legendary farmer ... tried 'one hundred kinds'[1] of grasses (or plants) and therefore [the traditional Chinese] medicine began" (*Si Ku Chuan Shu*, pp. 11-40).[2] According to *Huai Nan Tze*, the legendary farmer tried various kinds of grasses (plants) to learn about the nature and function of each so that people could either avoid them or make use of them. In fact, on one day he was poisoned seventy times by seventy kinds of grasses (plants) (*Huai Nan Tze*, 1994, p. 958). In modern terms, the legendary farmer performed a medical self-experiment by trying grasses (plants) in order to gain accurate knowledge of the medical function of various plants for the sake of the general public. In doing so, he made himself a human subject of that experiment.

What is interesting for us here is not only the fact that, in traditional Confucian China, medical experiments that used human persons as subjects existed, but also, given the Confucian understanding of the importance of knowledge to the public good along with the tentative feature of medical knowledge, the Confucian attitude toward the development of medical knowledge. It is this Confucian attitude and its related ethical assumptions that I would like to explore below.

As it is evident, the legendary farmer was motivated by the thought that the knowledge of the medical functions of various kinds of grasses was important for the health of the general public, and that if this knowledge could be obtained only through experimenting with human subjects, such experiments ought to be conducted. And as it is, his action is sanctified by Confucianism. From a Confucian perspective, what society cannot afford is to be ignorant and, thus, lacking the knowledge to cure the sick, to relieve the suffering, to save life, etc. A society thus cannot afford not to take medicine seriously, not to take having medical and biological knowledge seriously, for medicine and knowledge of medicine are necessary conditions in improving the health of the people, the quality of life in a society, and, furthermore, in allowing for the condition of human flourishing. It is not just the spirit of making a sacrifice for the sake of the public good, but also the devotion to the development of medical knowledge which makes the legendary farmer's self-trial a highly praised exemplary practice.

To understand the Confucian attitude toward knowledge, we need to understand the Confucian view on the inseparability of knowledge and humanity. The view is characteristically advocated by Confucius himself in his remark: "without knowledge, there can not be humanity" (*The Analects*, 5:19). Of course, in this remark, by "knowledge" Confucius means not only knowledge of some particular field, such as medical knowledge, but also practical wisdom or ethical knowledge of us as human persons. Nonetheless, the relevant point of this remark remains clear: while humanity is the guiding principle for life, to realize humanity requires knowledge of various practical fields (e.g., medicine, biology). For example, humanity requires us to help others in eliminating disease, deformity, and premature death, in releasing the suffering of patients, in improving the health environment and condition of the general public, etc. Yet to accomplish all these requires medical and biological knowledge. Therefore, from a Confucian perspective, a doctor or physician would not be able to realize humanity if s/he did not have the knowledge to help others to do away disease, deformity, premature death, etc. In other words, a lack of knowledge makes it impossible for a doctor to practice and realize humanity.

It is proper to mention in this context that in Confucian China, the word "*ru*," meaning Confucian, has also a connotation of "being learned." A "*ru*" person is someone who is learned. In Confucian culture, being "*ru*" and being "*ren*" (benevolence) both are ethical requirements for doctors, physicians, and pharmacists. A physician or doctor who is ignorant and incompetent is as bad as one who cares about nothing but profit, money, or other form of personal gain. Thus, Gong TingXian, a 16th century physician of the Ming dynasty put forth his famous "ten commandments" for doctors and physicians based on the ideas of being humanistic, righteous, and knowledgeable.[3] For Gong, to have medical knowledge is as important as it is to have senses of humanity and righteousness. And Gong's ten Confucian commandments are respected as ethical norms in China even today. At any rate, an insistence on the inseparability of knowledge and humanity, entailing that a physician or doctor must be both humane and knowledgeable, is a defining feature of what I take to be Confucian medical ethics.

Confucianism emphasizes not only the importance of knowledge, but also its tentativeness . For traditional Confucianism, to have knowledge, one must investigate the object of knowledge or *Gewu*. As *The Great Learning* says: "the extension of knowledge consists in the investigation of things" (Chan 1971, p. 86).[4] True, an individual might acquire a specific piece of knowledge through indirect channels; i.e., one can acquire a piece of knowledge by reading books or learning from other persons' experiences. However, if no one in the world ever directly investigates things, then there would never be any real knowledge of anything in the world. With regard to the extension of medical and biological knowledge, there are circumstances in which the investigation of things requires experiments with human subjects, e.g., the case of the legendary farmer, which is an example of a justified experiment of this type. Of course, Confucianism will not say indiscriminately that all medical and biological experiments with human subjects can be justified as long as they are oriented to obtain medical or biological knowledge. For example, Confucianism will categorically reject experiments (with human subjects) that are meant ultimately to develop biological or biochemical weapons, though taking a form of seeking new knowledge. Nonetheless, Confucianism does see that in some circumstances, experiments with human subjects are both practically necessary and ethically justified. The concrete circumstances and conditions in which Confucianism would consider an experiment to be justified will be discussed in details below. Suffice is it here to point out that an insistence on the extension of knowledge, together with an insistence on the extension of humanity, is one of the defining features of Confucianism. In addition, this Confucian emphasis on the extension of knowledge also leads to a Confucian conception of the tentative feature of human and medical knowledge and thus to a Confucian experimental spirit in seeking knowledge.

A twofold moral can be drawn here. On the one hand, medical knowledge and its expansion is taken by Confucianism to be necessary and indispensable for human flourishing and well being. On the other hand, just as Confucianism would not indiscriminately sanctify a practice as long as it is meant to develop new knowledge (e.g., the practice to develop biological weapons), it would not indiscriminately sanctify all kinds of medical and biological experiments with human subjects without a critical examination of an experiment's nature, purpose, procedure, and result. Then the question for us here is: What kind of medical experiments with human subjects will Confucianism sanctify? What conditions should they meet?

To illustrate the problem further, let us consider the following case. Three prisoners who are scheduled to be executed in two days are stripped of all their political rights by their death sentence. Now there is a medical experiment with a new kind of anti-cancer drug that needs human subjects, and naturally it is proposed that these three prisoners should be used as subjects in the experiment. Does Confucianism sanctify this kind of experiment? If so, would it support the experiment without any condition or would it insist on some pre-required conditions? Let me clarify the second question in the manner of making a comparison of this case to the case of the legendary farmer. First, because a prisoner is not the person who conducts the experiment but merely the person who is being experimented on, it raises the question about whether the prisoner's consent should be consulted before he is made a subject in the experiment. In his self-experiment, the decision of the legendary farmer to experiment on himself presupposes his consent to the experiment. In short, is voluntary consent necessary for a medical or biological experiment from the Confucian perspective? Second, since the prisoner is not a regular experimentee but a special one, more specifically he is a person who is sentenced to death and stripped of all political rights by the court, should this make any difference on the issue concerning the prisoner's consent from a Confucian perspective?

To attend to these questions, let us now explore the Confucian justification of medical and biological experiments with human subjects and the rationality embodied in this Confucian mode of justification.

II. A CONFUCIAN JUSTIFICATION OF EXPERIMENTING WITH HUMAN SUBJECTS

Because Confucianism normally evaluates medical and biological experiments with human subjects in terms of their possible contribution to the extension of knowledge and, thus, its contribution to humanity and social progress, one might intend to take this form of justification as a kind of bioethical utilitarianism. This identification is, in my judgment, incorrect. To see this, let me first try to outline some basic characteristics of a Confucian justification.

I will start with a Confucian account of a justified therapeutic medical experiment with human subjects. A Confucian account might go something like this:

> A therapeutic medical experiment with human subjects is justified if its performance under the circumstance would be supported strongly by scientific evidence as leading to more effective treatment of *the subjects as patients* than its alternative medicine or procedure of treatment, and if its performance under the circumstance would not be disallowed by the principle of righteousness in line with humanity.

The term "strongly" suggests that the first requirement does not demand absolute certainty of the result or effect of an experiment because there is always some uncertainty in conducting an experiment aimed at developing a new treatment or medicine better than its alternative. Also, in the traditional Confucian medical community, there is a belief that no two cases of sickness are exactly the same; thus, any therapeutic intervention involves a certain degree of uncertainty. The question is to what extent is uncertainty allowed. On the other hand, an experiment must not be conducted if the physicians or scientists do not have good scientific grounds to believe that the new treatment or medicine would be more effective than the existing treatment or medicine. This requirement rules out any unnecessary experiment that lacks scientific ground and a perceivable practical merit, and thus exposes patients to unnecessary risk and discomfort.

"Righteousness in line with humanity" is the principle that (1) "one should wish the good for the other as the same that one wishes for oneself" (*The Analects*, 6:30) and, accordingly, "one does not wish the other to force something on oneself, therefore, one must not force some thing on the other" (*Ibid.*, 5:12), "one must not impose upon the other *that which* one does not desire for oneself" (*Ibid.*, 12:2) and that (2) human commiseration, compassion, and care about the other as a fellow human being and, thus, a sense of being shameful if one does wrong to the other (*Mencius*, 6A: 6-7). The condition "would not be disallowed by the principle of righteousness in line with humanity" is to rule out experiments that are performed under circumstances contrary to humanity, e.g., using unjustified coercion.

In short, a Confucian justification of a therapeutic experiment consists in the argument that an experiment is justified when it is a wise thing to do based on the scientific evidence and, in addition, a right thing to do because it confirms the spirit of righteousness in line with humanity, care, and propriety. Accordingly, an experiment should be rejected if it is either not scientifically wise or it will do wrong to the patient, or both. Good reason here thus consists not only in scientific evidence, but also in the idea of righteousness. Therefore, Confucianism will sanctify all therapeutic experiments that are patient-centered, meaning experiments that aim to serve the patient, and that are performed with good scientific grounds while meeting certain conditions required by the principle of righteousness in line with humanity. And the therapeutic experiments sanctified by Confucianism will include, as I understand it, experiments with infants. That is, as long as such an experiment is meant to do good to the infant, is performed by doctors or physicians with expertise and competence, and has the informed consent of the parents of the child, it will be endorsed by Confucianism.

I shall now turn to non-therapeutic medical and biological experiments. To give an example of what I mean by a Confucian justification of non-therapeutic experiments with human subjects, a Confucianist account of the *justifiability* of an experiment of this kind might be stated as follows:

> A non-therapeutic medical or biological experiment with human subjects is justified when its performance under the circumstances would be vouched for by the possibility that the knowledge it yields could have a significant contribution to the welfare of a community in particular or the welfare of humankind in general, a contribution that no one could reasonably reject or deny, and, in addition, when its performance would not be disallowed by the principle of righteousness in line with humanity.

This is intended as a characterization of the features of the kind of non-therapeutic medical or biological experiment that Confucianism would see as justified. This characterization is only an approximation, which will need to be modified as the contexts in which experiments occur vary. Here I would like to offer a few clarifying remarks.

The idea of a vouched for contribution that one cannot reasonably deny or reject is meant both to exclude those experiments that do not contribute anything (i.e., no new knowledge, no new procedure or method of treatment and curing, etc.) to the welfare (i.e., the health) of the community or humankind and to insist that such a contribution is beyond a reasonable doubt. The term "vouched for" implies that a perceived contribution enjoys the support of strong scientific evidence and its picture, as it is perceived, is reasonable to those who are involved in the experiment. In addition, what Confucianism demands here is not just a contribution, but a significant contribution. That is, such a contribution must be important to the extent that the burden which individual subjects must bear for the sake of this contribution can be reasonably accepted as a kind of necessary, worth-paying *price*. Thus, such a contribution must be significant to the extent that no one can reasonably deny it and disregard the necessity of using human subjects.

The idea of "not being disallowed by the principle of righteousness in line with humanity" is meant not only to rule out inhumane experiments, but also to exclude experiments under inhumane circumstances. Inhumane experiments refer to those experiments that are not meant to serve but to destroy humanity. An example of this kind of experiment is that which is meant to produce bio-chemical weapons, bacteria weapons, etc. Experiments that are performed under inhumane conditions include those that coerce specific groups to become subjects of the experiment, those that do not make necessary preparation to reduce the burden and suffering of the human individuals (the subjects) in the experiments, those that are not based on solid scientific knowledge and evidence but are performed because of the availability of subjects in specific circumstances (e.g., there are plenty of subjects if one were to use war prisoners or prisoned criminals, etc.) or due to political expedience, those that are both deceitful and exploitative, etc.

Now a distinction between Confucian and utilitarian justifications is in order. A utilitarian justification argues that the social benefits to be gained from experimenting with human subjects outweigh the harm that is caused to the individuals who are treated as "subjects" in these experiments. It also argues that the possible harm from not performing a particular experiment with human subjects may be far worse than the harm caused to the society or to the individuals who are treated as "subjects." To put it in another way, the utilitarian justification consists of two parts: the first part is the thesis that the benefits gained from the experiment outweigh the harm, and the second is that the society should always choose the lesser of two harms. Certainly there are some convergent areas between Confucianism and utilitarianism with regard to the justification of medical and biological experiments with human subjects. However, these convergent areas must not obscure the important differences between them.

First, although both Confucianism and utilitarianism emphasize the possible good to the general public (or to the individual subjects involved) resulting from an experiment as a reason to justify the use of human subjects, Confucianism's conception of the public good has moral content, while utilitarianism mainly emphasizes the quantity of good that is produced. Confucianism is not arguing that a greater number of practical benefits is sufficient to justify an experiment whose cost (i.e., the harms and burdens) is outweighed by the utility of these benefits. Rather Confucianism is arguing that in general knowledge is important and indispensable and at times requires a price, e.g., experiments using human subjects. Therefore, the question that utilitarianism asks is first and always how much good (practical benefits) an experiment will produce and whether or not the potential gain resulting from the experiment outweighs the harm that is caused to individual subjects and the burden that each subject must bear. The question of Confucianism is first and always what kind of good an experiment will produce. To see the difference between these theories, let us consider the case of the legendary farmer. A utilitarian would read that the legendary farmer's self-trial produced a greater sum of good that far outweighed the harm that was done to the legendary farmer and the burden that he bore; this is the most important thing that we must attend to. A Confucian would read it as something that goes like this: on the one hand, we must appreciate that the sum of the good produced in the farmer's self-trial is far greater than the harm incurred by the experiment. On the other hand, and of greater importance, we must understand that humankind needs medical knowledge, and the practical piece of medical knowledge that the legendary farmer meant to produce could be gained only through a medical trial with a human subject.

Second, in the utilitarian justification the idea of righteousness as a criterion of justification is absent while in Confucianism this idea is at the very core of justification. An absence of the idea of righteousness in turn seriously handicaps utilitarianism, making it unable to emphasize a distinction between right and wrong benefit and revealing a serious inadequacy in dealing with persons as humans in any circumstance. On the contrary, incorporating the idea of righteousness not only gives

Confucianism a solid normative conceptual resource, but also enables it, more than utilitarianism and any other ethical justification of medical or biological experiments with human subjects, to attend to the essential issue: a defense of humanity. The distinction between Confucianism and utilitarianism can be seen from the above example of the experiment with the three prisoners scheduled to be executed. From a utilitarian stance, it seems that there is no need to ask for the voluntary, informed consent from these prisoners for they will be executed and it might be more beneficial to the general public if they are used as subjects in a medical experiment. To be sure, utilitarianism might not object that these prisoners are consulted. What utilitarianism rejects is that they must be, ought to be, consulted. Confucianism, on the other hand, will hold that these prisoners should and ought to be consulted. Confucianism will insist, as Tung ChungChu, a Confucianist of the Han dynasty, put it, that a society "should make straight the righteousness but not pursue benefit at the cost of righteousness." In addition, Confucianism will argue that since no medical researcher would wish to be forced or misled into being a human subject of an experiment, to bear the suffering and burden without consenting, researchers of a medical or biological experiment should acquire the informed, unforced consent from the three prisoners.

Third, utilitarianism cannot account for some crucial requirements for a justified experiment, e.g., why informed, unforced agreement from an individual subject is indispensable. If utilitarianism appealed to the idea of individual rights, then it would undermine its thesis that the greater amount of good is a necessary and sufficient condition for the justification of a medical or biological experiment as it would show that a stronger reason than the greatest sum of benefit is required in order for an experiment to be justified. If it stays with the belief that a greater amount of good outweighs a smaller amount of harm, then it cannot explain why in such a context informed, unforced agreement of individual subjects to bear the burden of small harm for the sake of greater amount of public good would be necessary. By appealing to the idea of righteousness in line with humanity, Confucianism can give an adequate account of the indispensability of informed, unforced agreement of individual subjects, which I will discuss in detail below, and other necessary requirements for a justified experiment. In addition, utilitarianism cannot account for or appreciate the moral significance of some contingent facts involved in an experiment. Taking the case of the three prisoners mentioned above, suppose that one prisoner sincerely regrets the crime that he had committed and thus voluntarily wants to be a subject in an experiment in order to redeem his crime. Confucianism would interpret the act of this prisoner as an act of an awaking person, namely a person whose human nature was sleeping and now is aroused, and thus would sanctify the act. Utilitarianism, on the other hand, would not see a difference if this fact does not change the end result of the experiment.

Finally, utilitarianism might not be able to account for why certain kinds of experiments might be rejected. Consider, for example, experimenting with infants. Utilitarianism might not reject such an experiment if it leads to the greatest practical benefit. Confucianism, on the contrary, would generally reject such an experiment on the ground that making a child bear the burden for others is an act that completely lacks compassion, and is an act of cruelty. This is true even if the child's parents consent to their child to being a subject for Confucianism will argue that the child's parents have no love for their child, and such a lack of love is wrong because it indicates that the parents' human nature is lost since any authentic love for their child would lead them to refuse to agree to make their child a subject in an experiment. In either case, the infant child should not be a subject in an experiment and could become a subject only for wrong reasons. Therefore, there are some significant, even essential, differences between utilitarianism and Confucianism concerning the ethical justification of medical or biological experiments with human subjects. Utilitarian justification is purely practical, existential, and its sole criterion is the sum of good (practical benefit) that an experiment will produce. This is not an exception to so- called rule utilitarianism, for it is one of the defining features of rule utilitarianism that the rule which an experiment should abide by is the rule of the greater amount of good. On the other hand, Confucianism, while emphasizing the possibly important practical contribution when it sanctifies an experiment, stresses the fact that the experiment is operating with the scheme of righteousness in line with humanity, both in purpose and in procedure. Otherwise, even if an experiment might produce greater good, Confucianism would still reject it. For example, suppose the three prisoners are forced to be subjects in the medical experiment described above and are deliberately abused. Even if such an experiment might indeed produce a greater good than if the three prisoners were merely executed, Confucianism would still denounce the experiment as inhumane and reject it.

Confucianism does not rule out the possibility that there is a universal justification for medical and biological experiments with human subjects. For it is at the core of Confucianism that an ethical justification of one type of human practice should confirm the principle of *dao yi*, the righteousness of the *dao*. Moral reasoning is important here to the extent that it is through moral reasoning that human agents arrive at *dao yi*. In the context of bioethics, and in particular, the experimentation with human subjects, the justification of medical and biological experiments ultimately appeals to the *dao yi* that is not provincial, but universal.

However, Confucianism does insist that the ethical justification of a medical or biological experiment with human subjects is characteristically historical, practical, and, thus, *perspectival*. That is, justification is always made by human agents historically situated. Human agents can only justify their practice from their historical understanding of *dao*. In addition, such a historical understanding itself is mediated by practice and always embedded in the latter. This means that a historical justification itself is also a practical justification. Therefore, though *dao* is the ultimate arbiter of right and wrong, human agents can only think of *dao* from their historical and practical situation. It is in this sense that Confucius said, "it is humans who can make the *dao* great, not the *dao* that can make humans great" (*The Analects*, 15:29). It is humans who interpret and practice the *dao* historically, not the *dao* acting transcendentally from "above" to make humans and their practice great. Since ethical and moral justification is historical and practical, it is perspectival, meaning that it is from a particular intellectual and moral world view that is developed in a historically evolving community of social practice. This indicates that it neither excludes the possibility that there is an alternative interpretation of the *dao* or a justification of some practice, e.g., medical and biological experiments with human subjects, nor does it preclude a specific justification from making a universal claim. Instead, what it emphasizes is that a specific justification should be aware that its universal claim is always made from a specific perspective and thus needs to open itself to further scrutiny, modification, and further development. Thus in Confucianism, justification is *characteristically*, rather than *essentially*, historical, practical, and perspectival. And universal *dao yi* is immanent in these justifiable claims.

In short, so far as the justification of a medical or biological experiment with human subjects is concerned, Confucianism appeals to the possible significant contribution of an experiment to the individual subjects, to the community, to humankind in general, to possible social progress, and to the principle of *dao yi* embodied in the idea of righteousness in line with humanity. It sees that in certain circumstances, the practical necessity of using human subjects in a medical or biological experiment, the demands of humankind or social progress, and righteousness in line with humanity, merge together. In turn, the point at which such a merge occurs constitutes the point of departure for the Confucian justification of a medical or biological experiment with human subjects. Thus, the Confucian justification, while sharing certain views with the utilitarian justification, differs fundamentally from the latter. Utilitarianism is concerned with the idea of the greater material gain or loss. Confucianism transcends utilitarianism with the idea of righteousness in line with humanity. An absence of the concept of righteousness handicaps utilitarianism, rendering it unable to account for some crucial aspects of justified experiments and some requirements for a justified procedure. An insistence on righteousness in line with humanity empowers Confucianism. For example, it offers a solid normative reason why certain requirements (e.g., the informed, unforced consent of individual subjects) for a justified procedure of an experiment are necessary. Now I shall continue the discussion of the Confucian justification by exploring a Confucian view on informed, unforced consent of individuals subjects in a medical or biological experiment, and thus demonstrate why I prefer Confucianism to utilitarianism.

III. RIGHTEOUSNESS, INFORMED-UNFORCED CONSENT, AND HUMAN DIGNITY

1. Righteousness and informed-unforced consent

As discussed above, inherent in the Confucian justification of a medical or biological experiment with human subjects is the concept of righteousness in line with humanity. This leads Confucianism to insist that medical and biological experimenters must not force or mislead individuals into becoming subjects in an experiment and treat them as mere means to specific ends. In other words, for Confucianism, informed, uncoerced consent of the individuals to become *subjects* in an experiment is essential and indispensable for a justified procedure. That an informed, uncoerced consent of individual subjects to participate in an experiment is inscribed in the idea of righteousness in line with humanity can be seen in the following three points.

First, as mentioned above, the principle of righteousness in line with humanity dictates that one must not do to the other that which one does not wish for oneself, and one must help the other with establishing that which one wishes for oneself. Since an experimenter, like everyone else, does not wish to be misled or forced to be a subject in a medical or biological experiment, s/he should not mislead or coercively solicit the other into being a human subject in a medical or biological experiment. Since an experimenter would want to be well informed and able voluntarily to consent or decline being a human subject in a medical or biological experiment, s/he therefore ought to inform the individuals who are to be subjects and allow them voluntarily to consent to or decline participation in the experiment.

Second, as described above, inscribed in the Confucian principle of righteousness in line with humanity are the ideas of compassion and benevolence, which in turn require that an experiment and its executor care for the individual subjects who will bear certain burdens in the experiment. One way to demonstrate or safeguard such a compassion and care is to understand them as human persons who have feelings, to empathize with their sufferings and burdens, and, therefore, to see that informed, uncoerced consent is to be sought for just these reasons. Accordingly, misleading or forcing them to be subjects is wrong for just these reasons. This should be true even in the case of special subjects such as prisoners or other individuals who are not regular citizens. That is, from the perspective of righteousness in line with humanity as described above, when particular irregular individuals (e.g., prisoners) are employed as subjects in a medical or biological experiment, these individuals should be informed and their voluntary consent should be obtained.

Third, in Confucianism, experimenters are trusted that they will inform and obtain voluntary consent from the subjects in their experiments. Indeed, a particular feature of the Confucian culture is a trust that a physician or doctor will not forget what is right in favor of some benefit or practical expedience. When they do otherwise, they betray this trust and violate righteousness in line with humanity in an utmost serious manner. In this context, as Confucius insisted, "trust is approximate to righteousness" (*The Analects*, 1:13). Because of this trust experimenters should do their best to help individual subjects understand what is involved, especially the purpose and conditions of the experiment, as well as the possible risks. If they do otherwise it is not just a matter of a failure to meet expectation, but also a betrayal of trust.

Therefore, for Confucianism, informed, uncoerced consent of individual subjects is inscribed in the idea of righteousness in line with humanity in the context of medical and biological experiments. It rules out coercion and the deliberate misleading and disrespecting of the individual participants (subjects). Here, the pressure of the idea of righteousness in line with humanity on a medical or biological experiment is always the question: Does such a practice treat the individuals involved as human persons? Does each individual have the conditions and the resources to act as a human person? Therefore, while the idea of righteousness in line with humanity constitutes one of the most effective conceptual resources for the prevention of the impairment of humanity in medical and biological experiments, informed, uncoerced consent is a necessary policy to materialize and actualize the idea of righteousness in a medical or biological experiment. Now I shall turn to the issue of righteousness and individual rights.

2. Righteousness and individual Rights

As indicated above, Confucianism takes the dignity and sacrosanctity of human persons seriously and thus argues for a humanistic procedure for medical and biological experiments, at the core of which is compassion, the benevolence of the experimenters, and the informed and unforced consent from the individual subjects. So far as the insistence on informed and unforced consent of individual subjects is concerned, Confucianism shares this view with liberalism. However, as discussed above, Confucianism does not invoke the idea of individuals' rights when arguing for the necessity of informed, unforced consent. Rather Confucianism appeals to the idea of righteousness in line with humanity. Thus, Confucianism also differs from liberalism.

I believe that the Confucian argument has its strength. If it is because of the requirement of natural rights for life and happiness that there must be voluntarily consent from the individual subjects, then the first question is why consent must be informed consent. That is, if it is just a matter of rights to choice and action, then nothing here implies an *obligation* of the experimenters to inform the subjects. One could argue that you have a right to a voluntary and informed choice, but it does not follow that I have an obligation to tell you what you should know. (Just as you have a right to have a wife, it does not follow that I have an obligation to be your wife or help you with finding a wife.) If one is considered to have an obligation because of his/her role as an experimenter, and the rules or laws regarding the procedure of an experiment require this obligation, then the obligation arises, at least directly, from the experiment's social role and these particular rules or laws, not from the rights of individual subjects. In addition, if the obligation arises from relative, administrative rules or laws concerning medical or biological experiments, then this legal obligation should be distinguished from moral obligation. Thus it is righteousness in line with humanity, rather than individual rights, on which we should turn in arguing for the informed, unforced consent.

Second, the strength of the Confucian approach can also be seen if we consider the case of the three prisoners mentioned above. If it is the rights of a person that require an experimenter to inform these prisoners, one might respond: "But they are stripped of their political rights and sentenced to death. They are not regular persons who enjoy regular rights, and thus we need not to be bothered by the issue of violating their rights." Of course, I am not saying that the rights-argument is incorrect. What I am saying is that it is weaker than the Confucian argument.

Third, another question to be put to the rights-argument is: Does an individual have a right to consent to do something that might threaten his/her personal life and happiness? And, perhaps, on this point, it makes even less sense to speak of individual rights in this context. For example, it makes sense to ask, "Am I right or wrong to agree to be a subject in an experiment?" But it does not make sense to ask, "Am I defending my rights or violating my rights to agree to be a subject in an experiment?" or, "Do I have a right or do I not have a right to agree to be a subject?" (The second question makes less sense as it seems to be relatively irrelevant here.)

Fourth, the same is true of the procedure of an experiment when experimenters make certain professional decisions, e.g., to divide the subjects into different sample groups for the purpose of research and investigation. It makes sense for the experimenters to ask, "Is such a division necessary and right thing to do?" It makes less sense for them to ask, "Does such a division respect individual rights by obtaining consent from the individual, or does such a division infringe on individual rights in an unjustified manner?" and even lesser sense to ask, "Does such a division infringe on individual rights?" It makes more sense for researchers to ask, "Is it right or wrong to solicit human persons as subjects in an experiment?" But it makes less sense to ask, "Does informed, unforced consent from the individual justify an infringement or violation of what is inviolable in experiments with human subjects?" Thus, it is righteousness in line with humanity, not the natural rights of individuals, that is turned to and emphasized in the context of experimenting with human subjects when Confucianism insists on the informed, unforced consent from individual subjects.

Fifth, the inadequacy or problem of appealing to individual rights is also exhibited in other contexts. For example, in a Confucian culture, it is accepted practice for parents to make decisions on behalf of their infant child. This is not because in a Confucian culture it is considered that parents have rights over their child. Rather it is that in a Confucian culture it is considered right and wise that parents should make the decisions because, obviously, parents offer a more mature judgment; in addition, parents should and can make the decisions on their infant child's behalf on the basis of their love and care for their child's well being. Furthermore, parents love and care for the well being of their child is considered to be inscribed in the human nature of the parents (when they lose this love, they lose a part of their nature).

Of course, what is said above does not imply that individual rights should be ignored or not respected in medical and biological experiments with human subjects. Indeed, as I shall understand it, respect for individual rights is one of the moral sources for informed and unforced consent of individual subjects. What is argued above demonstrates the ideas of the individual rights of the subjects is still an outer moral resource–a source from without, not within, the moral agents–for the experimenters when it comes to obtaining consent from individual subjects. Other than this outer source, more important is an inner moral source. This inner source is the moral agent's sensibility of righteousness in line with humanity. From a rhetorical point of view, righteousness in line with humanity also gives Confucianism a stronger first premise of moral reasoning than is the idea of individual rights, in particular with regard to informed, unforced consent of individual subjects and other crucial matters of the procedure of an experiment.

At any rate, while the idea of righteousness in line with humanity is not conflicting but intersecting with the idea of individual rights, the former is what Confucianism appeals to when it comes to argue for informed, unforced consent of individual subjects. Indeed, when arguing for or against other requirements for a justified experiment with human subjects, the idea of righteousness in line with humanity is also that to which Confucianism appeals. I believe that the Confucian approach thus has its unique strength because it argues for some necessary conditions and requirements (e.g., informed, unforced consent) for a justified procedure of an experiment from a broader humanist perspective. Now I shall turn to the issue about *jing* and *chuan* in Confucianism.

3. A further consideration of righteousness: Jing[c] *and* quan

Hitherto, we have left the crucial Confucian thought of *jing*[c] and *quan* out of our discussion. *Jing*[c] literally means "the basic principle." *Quan* means "flexibility," making adjustments to a given practical context when one applies a basic principle to this context. An example of the Confucian idea of *quan* is Mencius' argument that though the principle of propriety requires that a man and woman (who are not a married couple) should not touch each other physically, a man would be a wolf if he does not give his hand to his sister-in-law and rescue her when she is drowning in water (*Mencius*, 7B:17). Here, a man giving a hand to his drowning sister-in-law is not a violation of the basic principle, but a policy of *quan* in this given situation.

The importance of *quan* in Confucianism can be seen in Confucius' worry that his followers might be able to be with him so far as abiding by the *tao* is concerned, but be unable to follow him when it comes to *quan* (*The Analects*, 9:30). It can also be seen in Mencius' characterization of the inability to be *quan* in understanding righteousness as practicing bandits' ways (*Mencius*., 7A:26). Thus, one would not really understand a Confucian view on righteousness in line with humanity in medical and biological experiments without an understanding of the Confucian idea of *quan*. While to abide by the principle of righteousness is an important requirement for a justified medical or biological experiment with human subjects, proper interpretation and flexible application of the principle, or an ability of *quan,* is critically important for practice.

Without the concept of *quan*, a dogmatic appeal to the idea of righteousness in line with humanity would be misleading. Consider the following case of a patient who is totally incompetent in making judgments. Now suppose this patient's doctors believe with good scientific grounds that a new kind of anti-cancer medicine might be effective in treating his cancer. According to Confucianism, it would not make much sense to insist that the physicians should first obtain his consent before they can justifiably give him the treatment that they believe will do him good. But, as we discussed above, a ramification of the idea of righteousness in line with humanity is the requirement for informed, unforced consent from individual subjects. In this case the patient is not able to offer competent consent. How should we solve such a dilemma? In this situation, Confucianism would say that what is needed is *quan* or *bian tung*, literally meaning "making change." This does not deny the general idea of the necessity and indispensability of informed, unforced consent of the patient, rather it is to stress a flexible application of this idea in this abnormal situation. Of course, to prevent physicians or doctors from abusing their authority and position it might be prudent for an additional requirement, e.g., consent of a relative or family member of the patient, to be imposed. However, *quan* or *bian tung* is needed, implying that we must not be too dogmatic in insisting on the informed, unforced consent of the patient here. What matters here is that which is really good for the patient, and whether the action is righteous in spirit and in reality, that is, both intending the good and really doing the good. In this sense, *bian tung* does not mean to deny or undermine the idea of righteousness in line with humanity, but to practice it flexibly in concrete situations.

Quan or *bian tung* is important when it comes to the question to what extent an individual subject or his/her representative (e.g., the parents of an infant child) should be informed. A Confucian answer to this question is that one has to see the concrete circumstances in which each experiment is performed. The bottom line is that experimenters must not deliberately mislead the other; they should have human compassion for the individual subjects. Here, not to specify to an individual subject certain possible harms of an experiment about which even the experimenters themselves are not sure is one thing, but deliberately creating the false impression that there is no harm involved is quite another. Also, since in a Confucian culture there is a belief that no two cases of sickness are exactly the same (i.e., because of varying physical constitutions and different degrees at health that individual subjects have, a same specific stimulus of an experiment might mean different results to different patients), there is *de facto* a consensus in the Confucian medical community (i.e., in China) that for different patients, physicians or doctors should deliberately define what they should reveal to a patient who will be subject to a new therapeutic experiment.

In Confucianism, while to abide by the idea of righteousness in line with humanity is generally a moral imperative, to be *quan* in the application and interpretation of this principle in various contexts or circumstances of medical or biological experiments is a part of the practical wisdom essential to experimenters using human subjects. This brings us to the Confucian understanding of the practical authority and responsibility of physicians and scientists.

4. Authorities and responsibilities of physicians or scientists

Given what has been said above, Confucianism is willing to grant doctors, physicians, and scientists significant authority and decision-making power corresponding to their roles and practical responsibilities in medical or biological experiments involving human subjects. Accordingly, Confucianism also sees that doctors, physicians, and scientists have due moral duties and responsibilities. This is partially suggested in the discussion in above. Now, I would like to add a few points.

First, for Confucianism, to serve people and humanity should not be an optional goal, but a required commitment for doctors, physicians, and biological scientists. In other words, it is a moral duty for doctors, physicians, and scientists to commit themselves to serving people and humanity. Acquisition of new medical or biological knowledge is intended to further this general goal. This also means that the procedure used to acquire medical and biological knowledge must be in line with this general purpose. Within this framework, it is the doctors, physicians, and scientists who should decide what is the best course to take from a scientific point of view.

Second, and corresponding to the first point, abiding by the idea of righteousness in line with humanity should also not be merely an optional policy, but a required moral duty. This means that it is a moral duty and a practical responsibility for physicians and scientists to treat individual subjects fairly, with commiseration, compassion, love, and care within the scope of their responsibilities and authorities. This also means that physicians and scientists must be honest to themselves and to the human subjects. That is, they must not lie to themselves because whatever practical reason or expedience, to the patients or subjects on what is right and what is wrong, what is the better and what is the worst concerning what is going on in an experiment.

Third, physicians and scientists indeed have a leadership role to play regarding the authority and decision-making in the scientific part of a medical or biological experiment. This authority can and ought to be exercised when it comes to the issue of what is the better, or the best, course to take in an experiment. This also means that it is a moral duty of the doctor, physician, or biological scientist to insist on the better course from a scientific point of view. For example, if a new kind of method of treatment is considered beneficial to cancer patients from a scientific point of view but is rejected by a patient for some practical reasons, e.g., because of certain religion beliefs, it is a doctor's practical responsibility and moral duty to do his/her best to convince the patient to accept the treatment. Moreover, it is the doctor's duty to make decisions on how the new treatment should be administrated, without yielding such decision authority to other people (e.g., the member of the patient's family, or other non-medical authority). Thus, while it is the moral duty of doctors, physicians, and scientists to commit themselves to serving people and humanity, it is their authority to use their knowledge and techniques in a way that they believe to be the best from a scientific perspective and wisest from a practical perspective. It is also their moral duty to ensure that this knowledge is respected and honored.

Division of English Classics, Philosophy, and Communication
University of Texas at San Antonio
San Antonio, Texas, USA

NOTES

[1] As it is evident in the text that "One hundred" does not mean the number "one hundred"; rather it is a metaphor for many.

[2] For details, see Shi MaZhen, 1986.

[3] For details, see Gong TingXian.

[4] *The Great Learning* is one of the four Confucian classics (the other three are: *The Analects*, *Mencius*, and *The Doctrine of the Mean*). This essay uses Wing-Tsit Chan's translation.

REFERENCES

Chan, Wing-Tsin: 1971, *A Source Book of Chinese Philosophy*, Princeton University Press, Princeton.

Chen, Yiping (ed.): 1994, *Huai Nan Tze*, Guangdong People's Publishing House, Guangzhou, China.

Confucius: 1974, *Analects*, Beijing University Press, Beijing.

Gong, TingXian: n.d., 'Ten Maxims for Physicians,' in *Back To Life From A Myriad Of Sickness,* Wanli edition, China.

Mencius: 1922, *The Doctrine of Mencius*, JiaoXun (ed. with footnotes), Commercial Publishing House, Shanghai.

Shi, MaZhen: 1986, 'The Records of the History: The Appendix of the Three Emperors,' in *The Complete Works of WenYen Four Treasures,* Taiwan Commercial Publishing House, Taipei.

Weggle, Oskar: 1991, 'Between Marxism and meta-Confucianism: China and her way back to normality,' in *Confucianism and Modernization of China*, Sike Krieger and Rolf Trauzett (eds.), Hase Kohler, Mainz.

PART IV

JUST HEALTH CARE
AND THE CONFUCIAN TRADITION

QINGJIE WANG

THE CONFUCIAN FILIAL OBLIGATION AND CARE FOR AGED PARENTS

Some moral philosophers in the West hold that adult children do not have any more moral obligation to support, mentally or physically, their elderly parents than does any other person in the society, no matter how much sacrifice their parents made for them in the past or what kinds of misery their parents are presently suffering. This is so, they claim, because children do not ask to be brought into this world or to be adopted. In other words, children do not give their consent to the parent/child relationship in the first place. The traditional filial obligation of supporting and taking care of the aged is thus left as either the private responsibility of the elderly themselves or as a societal burden on the public.[1] For example, Norman Daniels argues that there is a "basic asymmetry between parental and the filial obligations" (Daniels, 1988, p.29). The parental obligation of caring for their young children, says Daniels, is a "self-imposed" duty, while the so-called children's obligation of caring for their aged parents is "non-self-imposed" and thus cannot be morally required.[2] In her famous essay, "What Do Grown Children Owe Their Parents," Jane English also claims that a favor done without it being requested or a voluntary sacrifice of one for another can only create "a friendly gesture" (Sommers and Sommers, 1993, pp. 758-765). It incurs neither an "owing" nor a moral obligation to reciprocate. Therefore, what an adult child ought to do for her parents should not result from the debts she owes for her parents' past services or favors. It comes rather from the existing friendship and love relation between her and her parents. Accordingly, "a filial obligation would only arise," says English, "from whatever love (s)he [the adult child] may still feel for them [her parents]."[3] The moral obligation stops whenever the friendship relation ends. Because we cannot always assume a friendship relation exists between a parent and his/her children, filial obligation is not a genuine moral obligation at all.

In what follows I shall argue against the Daniels/English thesis in light of the traditional Eastern Confucian view of the nature of filial obligation. I shall make a distinction between "moral duty" and "moral responsibility" and argue that adult children's filial obligation of taking

Ruiping Fan (ed.), Confucian Bioethics, 235–256.

care of and being respectful to their aged parents should not be understood as a moral responsibility but as a moral duty, which is, by its nature, not necessarily self-imposed. That is to say, it is not consensual, contractarian, and voluntarist but existential, communal, and historical.

I. IS THE PARENTS/CHILDREN RELATIONSHIP "ASYMMETRICAL?"

We may find two basic theses that underline the Daniels/English rejection of adult children's moral obligation of taking respectful care for their aged parents. First, it claims that there is no consensual commitment by children in the filial relation because one does not choose to be the son or the daughter of one's parents. Second, it claims that filial obligation, if it is to be a moral obligation, should be based on the voluntary consent of all moral agents involved.[4] In order to challenge the Daniels/English argument, I would like first to discuss both of these basic theses.

As we have seen, Daniels holds that there is no consensual commitment by children involved in the filial relation. Because of this, he calls the relationship between parents and children "asymmetrical." But is it really asymmetrical? I believe that there are at least two problems here. First, it might be true that children do not give their consent in establishing the parent/child relationship. But it is also true that they do not give their consent not to establish such a relationship. Therefore, whether a child gives consent to being brought into the world, or to establish a parent/child relationship, is actually a misleading question. A child as a moral agent in such a relationship does not exist, and even if such a child did exist, it would be impossible for the child either to give, or refuse to give, consent in the first place. Thus understood, the right question should not be whether a child **actually** gives consent, but whether a child **would** give consent if the child were a full-fledged human person.[5] If we agree that life itself is the most valuable thing we have in the world and that most children have a normal family life, i.e., they receive adequate care and love from their parents[6], we should say that a child would give consent to the filial relation.

The second problem comes from the meaning of the word "consent." Is it really the case that we do not "consent" to be the sons or daughters of our parents? I do not think that Daniels would deny the fact that most of us did give our consent in establishing the parents/child relationship when we were young. But the real question is whether a child, who does not fully understand the consequences of the moral responsibility caused by her "consent," should be responsible later for her "premature" childhood consent. We may make a distinction between a strong sense of "consent" and a weak sense of "consent" here. By the strong sense of consent I mean a positive, explicit request or agreement which is made by a full-fledged human person. By the weak sense of consent I mean either a passive consent, e.g., not challenging a request to do something or having something done to you, or a consent made by a human person who is not full-fledged yet. In light of this distinction, we may reformulate our question as to whether consent in the weak sense could create any degree of moral responsibility at all. In the case of the filial relation between children and parents, the question should be whether consent from a child, who is a partial or a potential moral person, can create at least partial moral responsibility. As we know, our everyday moral and legal practices lead us to give a positive answer to the above question. For example, suppose that, Peter, a 20 year old young man, set fire to a house for fun, and that the fire caused injuries to people and large property damages. Obviously, he should be blamed and probably punished because he is morally and even legally responsible for what he has done in practicing his free will or "consent." But how about if he were not 20 years old, but 15 or even 10 years old? Should he not be blamed and punished at all because he cannot be fully responsible for his action? No. He must be blamed and even punished because he should take at least partial responsibility for his consensual committing of the wrong. Maybe the blame and punishment for children or teenagers should not be as severe as that for adults, or full-fledged moral persons. However, it does not mean that they should not be given any blame or punishment or not take any moral responsibility at all.

I think that Daniels' failure in not seeing the real nature of "consent" by children in establishing the filial obligation consists in his taking children simply as some kind of abstract, isolated and unhistorical agents, rather than the real human beings who live their lives existentially, progressively, and historically. He misses the simple fact that giving consent, practicing autonomy, and thus taking moral responsibility is or involves a process of growing and learning. Yes, we are not born with the ability to give consent in a full moral sense, we do not receive this ability overnight. For some conveniences of our social and legal arrangements we set up a definite line for adulthood. However, it does not mean that one should be exempted from any moral responsibilities created by one's "consensual activities" before that line. Following a similar way of reasoning, because we at least partially "consent" to a filial relationship with our parents when they do many lovely things to us and make many sacrifices for us when we are young, we ought to fulfill a filial obligation, at least "partially," when we grow up.

II. CONSENT AND MORAL OBLIGATION

Even if we accept the thesis that for children there is no consensual commitment involved in the filial relationship, or a modified thesis that there is no full consensual commitment, we can still justify adult children's moral obligation of taking respectful care for their aged parents by challenging the legitimacy of Daniels/English second thesis. Compared with the first thesis, the second one, which says that filial obligation, if it is qualified to be a moral obligation, must be based on the voluntary consent of all moral agents involved, is clearly more fundamental, and thus needs a deeper discussion.

It is not very hard to see that the second thesis expresses a general moral principle which underlies not only Daniels/English argument but also some major accounts of the nature of moral obligation in the modern West. I call it the "principle of intentional consent." "Consent" is required because a moral action ought to be approved of by all the persons involved in the action. It is "intentional" because an agreement or an approval ought to be reached voluntarily and without any kind of outside coercion or deceit. Very clearly, this principle gets its power from Kant's concept of a person as potentially an autonomous, rational, and free agent.[7] That is to say, intentional consent is simply an exercise of one's autonomy and rationality. Therefore, as a free, rational, and autonomous moral agent, I am morally responsible only for the consequences of those actions which I have committed voluntarily, without any coercion and deceit. Otherwise I will not see myself behaving as a free and autonomous being. Living in modern society, it seems that few people can really deny the importance of the principle of intentional consent and that of the concept of autonomy in our consideration of the nature of morality. However, is it the absolute and exclusive grounding of morality? That is to ask, is there any limitation of that principle in our moral practice, especially when we consider filial morality in dealing with the relationship between adult children and their aged parents?

Let me try to answer the question by looking at the following example. When Fred, a strong man and a good swimmer[8], went by a swimming pool on his way home, he found a three year old child Sheila was drowning in a swimming pool with another young child John crying nearby. Does Fred have any moral obligation to jump into the pool to save Sheila? Most of us, I believe, would say "yes" according to our common moral sense. If Fred does not save Sheila without having any serious excuse, he would at least be seen as a mean person. What interests us in this example is not whether Fred ought to save Sheila but why Fred ought to try to save her. Obviously, Fred neither made a promise nor gave consent to a request from Sheila's parents or Sheila herself to save Sheila when she is in danger. However, not giving consent does not sufficiently exempt Fred from his moral obligation to save Sheila in such a situation. To me, what makes Fred morally obligated in this case is: (1) that Fred is a human being and Sheila is a human being; (2) that Fred is a good swimmer and John, the only other human being around, is young and not a good swimmer; and (3) that Fred happened to be around and was the only one who could save Sheila, etc. That is to say, it is the existential or factical "being" of Fred, Sheila, and John rather than Fred's intentional consent that is crucial in Fred's moral obligation to try to save Sheila.[9] Similar examples in our contemporary social and moral life can also be found in the cases such as the moral obligation of the present generation of human beings to protect the environment and to preserve some of the natural resources for future generations; a citizen's obligation to defend her home country; a man's obligation to yield to and to protect women and children in dangerous situations; an off-duty doctor's obligation to save a patient's life in an emergency situation; a healthy and normal person's obligation to yield to a disabled or handicapped person; a patient's obligation not to have physical contact with healthy persons if she knows that she has an infectious disease; a strong person's obligation not to take advantage of the weak; a rich person's obligation to help prevent the poor from starving; etc. All of these demonstrate that at least some of our commonly and ordinarily accepted and practiced moral obligations can be justified without being preconditioned by the mutual consent of the moral agents involved in the action. That is to say, they are, pace Daniels, "asymmetrical" rather than "symmetrical."

The examples and arguments above require us to think more deeply about the nature of moral obligation in general, as well as that of adult children's moral obligation to their aged parents in particular. What are the major features which make a person morally obligated to do something? What role does a person's intentional consent play in having a moral obligation?

In order to answer these questions better, I would like to call attention to the nature of our understanding of "ought" or "moral obligation." When we say "A ought (not) to do X," or "A is obligated (not) to do X," it seems to me that we often have a confusion between two types of "ought/obligation."[10] One type of "ought/obligation" is caused solely by the intentional consent of competent moral agents involved in the action, and I call this moral responsibility.[11] That is to say, competent moral agents should be morally responsible for the consequences caused by their consensual actions. For example, I ought not to ask my students to take a final written exam, if I promised that they could write a term paper instead. Obviously, as a professor, I would have the option of asking my students to take the final exam if I had not made that promise. But my consensual action of offering the promise makes me morally responsible to keep the promise. Compared with moral responsibility, moral duty is another type of "ought/obligation." It does not necessarily depend on the competent moral agent's intentional consent. It is rather determined mainly by what kind of existential situation a moral agent is in and what kind of social role she plays. For example, a normal and healthy person is obligated to yield to a handicapped person because the latter is handicapped. Similarly, a hostess is obligated to show her hospitality to her guests while a stranger is not.

Someone may argue that although many of our moral obligations are determined by different existential situations and social roles we play, we do often consent to be in those situations and to play those social roles in the first place. For example, no one forces me to choose to be a professor. But as soon as I have chosen, I am morally obligated to fulfil all the duties associated with the profession. My response to this argument is, first, we do not always choose our existential situations or social roles. Many times we are thrown into a situation and many social roles are imposed on us without our previous consent. For example, I did not choose to be a human being; I simply am a human being. And as a human being, I have certain moral duties such as not killing innocent human beings, keeping promises, taking care of myself, helping others, etc. A brother did not choose to be the brother of his sister, but nonetheless he is. And as the brother of his sister, he ought not to marry her or have a sexual relationship with her. Otherwise it would be incest which has been morally condemned by most cultural and moral traditions. Second, although many times a moral agent does theoretically have an option to play or not to play a specific social role, such an option may not always be practical and therefore not real. For example, how many Chinese or Americans are given a real chance not to be a Chinese or American, and thus are not obligated to be loyal to their homelands? Theoretically it is possible, but practically it is almost impossible. Therefore, it is implausible to claim that my moral obligation to be loyal to my homeland is based on my intentional consent rather than on the existential fact that I am a Chinese or an American. Third, consenting to do something and being obligated to do something are not always the same. In many cases they are different. For example, I may consent to lie to someone about something for certain reasons. But consent itself cannot make lying moral or morally obligatory. We need to ask further whether I ought to consent or not. Therefore, in many cases, I consent to do something because I ought to do it, rather than it being the case that I ought to do it because I have consented to do it.

Thus understood, moral responsibility and moral duty are two types of moral obligation. They are different and the distinction between them should not be confused. The difference, as I have argued above, consists in that the former is caused exclusively by the intentional consent of the moral agent while the latter is not. However, they are not totally different or completely irrelevant to each other. Moral responsibility may be seen as a special type of moral duty. That is to say, moral responsibility is a particular moral duty of a moral agent when she behaves as an autonomous being or when she practices her autonomy in her consensual actions. However, a human being as a moral agent is not only an individual autonomous being. She is also a social and communal being, which imposes on her duties for caring for others as well as for her surrounding natural environment, and a rational being, which makes her obligated to calculate the consequential implications of her consensual action before she consents to it. Furthermore she is also a historical and cultural being, a concrete and situational being, etc. All of these essential features of a human being have created or revealed different types of moral duties that human beings as moral agents have. Therefore, an appropriate moral evaluation or moral judgement of a person's action should be based on or determined by weighing these moral duties of the person in her existential situation against one another.

To illustrate this point let us return to our previous example of Fred's obligation to save Sheila from drowning. On the one hand, if Fred were not a good swimmer, or John were not a child but the father of Sheila and a good swimmer, Fred's moral obligation to save Sheila would be weakened, or even be released. On the other hand, if Fred had made a promise to save Sheila when she is in danger, his obligation would be enhanced. But in both of the cases, we should not say that Fred ought not, or that he is not obligated, to save Sheila.

In light of the distinction between the two kinds of moral obligation, i.e., moral responsibility and moral duty, it becomes clear that the filial obligation of adult children to take care of their aged parents belongs to the category of moral duty, which, by its nature, is existential rather than consensual. It is so because the family, which defines the adult children's filial obligation to their aged parents, is basically a natural community rather than a social contractarian community. As long as the natural family is still one of the basic forms of our social and communal life, the parental and filial obligations between parents and children will exist. Therefore, being a son or a daughter of one's parents, one is obligated or has a duty to respect them as parents and to take care of them if they are necessary, no matter whether one chose to be the son or the daughter of one's parents.

III. *XIAO* AS CARE FOR AGED PARENTS IN CONFUCIUS

We can see the existential nature of adult children's filial duty to take respectful care of their aged parents much clearer in the Confucian moral tradition. It is well known that Confucianism in general can be seen as a theoretical expression and a systematic justification of traditional family values in ancient China (Fung, 1948, p.21). *Xiao* (filial piety), which primarily defines children's moral duty to their parents, has been understood in the 2500 year old Confucian tradition as the "root" of morality (*Analects*, 1:2).[12] It is, in Max Weber's words, "the absolutely primary virtue" which "in case of conflict, ... preceded all other virtues" in China (Weber, 1951, p.157). We can find at least three important meanings of Confucius' term "*xiao*" in the *Analects*.[13] First, it means "*neng yang*," i.e., adult children's willing and being able to take respectful care of their aged parents (*Analects*, 1:7; 2:6; 2:7; 2:8; 4:9; 4:21; 17:21). Second, it means "*wu wei*," i.e., one's compliance with the way of, and never disobedience of the will of, one's father because father should be seen as a symbol of authority in the familial community and in the tradition (*Analects*, 1:11; 2:5; 4:18; 4:20; 13:18; 19:18). Third, it means "*shi yu you zheng*," i.e., to extend one's exercise of filial piety from family life to public and governmental service (*Analects*, 1:2; 2:20; 2:21). Very clearly, among these three meanings of Confucius' "*xiao*," the first meaning, i.e., taking respectful care of one's aged parents, is the most naturalistic and fundamental. The other two, which emphasize more the social and political applications of the term in traditional Chinese society, are extensions of the first meaning. Although these three meanings were often bound or mixed together in the Confucian understanding and uses of the term, there seems no logical necessity for the last two meanings to be derived from the first one. Therefore, in order to simplify my case, I would like to focus my discussion only on the first meaning of "*xiao*."

In the *Analects*, Confucius' term "*yang*" (care), as taking respectful care of one's aged parents, may be interpreted as: (1) taking care of one's parents' lives (*yang kou ti*); (2) taking care of one's parents' mental needs and making them happy (*yang zhi*); and (3) taking care of one's parents' spirits after they have died (*yang ling*).[14] On the one hand, Confucius thought that taking care of one's parents' lives is the fundamental, as well as the minimum, duty of a filial son to his parents. For example, he said that a filial son should "undertake the hard work when anything has to be done and let the elder enjoy wine and food when these are available" (*Analects*, 2:8). He also taught that a filial son ought to "try his best to serve his parents" (*Analects*, 1:7), to "worry about his parents' health" (*Analects*, 2:6) and "old age" (*Analects*, 4:21), and that he ought not to travel far away from his parents while they are alive (*Analects*, 4:19). On the other hand, Confucius emphasized that a true filial son ought not only to satisfy his parents' physical needs by serving them with fine food, clothes, and shelter, but more importantly to satisfy their mental needs and to enable them to live a happy life. Therefore, the care for parents must be accompanied by a genuine respect, which Confucius called "reverence" (*jing*[a]). In the *Analects*, when Confucius was asked once by his disciple about being filial, he said: "Nowadays for a man to be filial means no more than that he is able to provide his parents with food. Even dogs and horses are, in some way, provided with food. If a man shows no reverence, where is the difference?" (*Analects*, 2:7) In the *Mencius*, we are given examples of Zeng Shen's caring for his father Zeng Xi and of Zeng Yuan's caring for his father Zeng Shen to show the difference between care with full respect and that without full respect (*Mencius*, 4A:19). Moreover, Confucius thought that taking care of the parents' spirits after they died, e.g., providing a dignified funeral, mourning for, and offering sacrifices to, the dead parents, should also belong to the minimum filial obligation of adult children (*Analects*, 3:12; 17:21; 19:17).[15]

Confucius' emphasis on "*xiao*," as adult children's taking respectful care for their aged parents, had a tremendous influence in shaping the Chinese understanding of the nature of morality. On the one hand, taking good care of one's parents is often seen as a cardinal virtue of a moral person (*jun zi*) and constitutive of being a good citizen. For example, in Confucius' point of view, one cannot be expected to be a good citizen without being a filial son first (*Analects*, 1:2; 1:6; 1:7; 2:20). Mencius said that the most important feature of "*ren*" (humanity) is "loving and caring for one's parents" (*Mencius*, 4A:27; 13:15). As Lin Yu-tang, a famous Chinese writer of this century, pointed out:

> The greatest regret a Chinese gentleman could have was the eternally lost opportunity of serving his old parents with medicine and soup on their deathbed, or not to be present when they died. ... This regret was expressed in two lines by a man who returned too late to his home, when his parents had already died: 'The tree desires repose, but the wind will not stop; the son desires to serve, but his parents are already gone' (Sommers and Sommers, 1993, p.753).

On the other hand, that all the parents and the elderly received good care from their children in the last years of their lives was taken in Chinese tradition as proof of a good society and a good government. For example, King Wen, the founder of the Zhou Dynasty and a sage king in Confucius' eyes, was praised for his policy of taking good care of the aged. We read the following stories in *Mencius:*

> Bo Yi[16] fled from the tyrant Zhou[b17] and settled on the edge of the North Sea. When he heard of the rise of King Wen he stirred and said, "Why not go back? I hear that Xi Bo[18] takes good care of the aged." Tai Gong[19] fled from the tyrant Zhou[b] and settled on the edge of the East Sea. When he heard of the rise of King Wen he stirred and said, "Why not go back? I hear that Xi Bo takes good care of the aged."

> When there is someone in the Empire who takes good care of the aged, men of humanity will look upon him as their refuge (*Mencius*, 7A:22).[20]

Because of this, Mencius said that in a good society "a son and a younger brother should be taught their obligation of taking good care of their aged parents. The people with grey hair should not be seen carrying burdens on the street" (*Mencius*, 1A:7). Otherwise it would be a matter of shame for the children of those elderly persons as well as for the government.

This Confucian tradition of seeing one's taking good care of one's aged parents as a moral duty has also been reflected in the Chinese laws from the beginning. For example, according to the laws of the Tang Dynasty, a man should be acquitted from fighting against another if his parents or grandparents were attacked by the other; and an official must resign from his official position and return home during the mourning period for his dead parents.[21] In the Qing Dynasty, a serious penalty could be reduced, and even a death penalty could be changed, if the criminal happened to be the only son in the household and his parents or grandparents would not be properly cared for if he were in prison for lifetime or executed.[22] Today, according to the Chinese Marriage Law, adult children's moral duty of taking respectful care of their aged parents is defined as:

> Children have an obligation to support and to assist their parents ... When children fail in such duty, parents who cannot work or have difficulty with their living have a right to demand alimony from their children.[23]

Obviously, taking respectful care of one's aged parents is one of the most important moral duties of an adult child in Confucian China as well as in all East Asian societies. It seems that very few want to deny this fact.[24] However, the questions remain whether this Confucian understanding of moral obligation can be well justified and whether the justification still makes sense in our everyday moral practices.

IV. CONFUCIAN JUSTIFICATIONS

What is the Confucian justification for adult children's moral duty to their aged parents? That is to ask, what makes it one of the most important moral duties in Confucian Ethics? In earlier times, the answer to this question was so obvious for almost all Confucians that one would never ask such a question of justification. For example, in the *Classics of Filial Piety*, one of the early Confucian canons, filial piety was described as "the unchanging truth of heaven, the unfailing equity of Earth and the universal practice of human being." In China, a saying such as "a truth of heaven" or "an equity of Earth" is the same as saying it is a principle which needs no justification or is beyond any justification at all. However, the fact that a moral principle needs no justification does not mean that it could not be justified, and the fact that a moral principle did not need a justification in the past does not mean that it does not need a justification today. As a matter of fact, we may find different ways of justifying filial obligation in the history of Confucian Ethics. In what follows I will try to re-formulate some of the major Confucian justifications.

The simplest justification may be seen from Confucius' theory of the "rectification of names." According to Confucius, a name not only has an epistemological function, i.e., reporting about or referring to something in reality, many names are operative as well. That is to say, they carry within themselves some imbedded normative request for action. For example, names such as "father," "son," "prince," "minister," etc., not only report about the bare biological or social/political facts, they are also associated with the obligatory "norms" of being the father, the son, the prince, and the minister.[25] Therefore, saying the word "son," for example, is not primarily pointing to a referential entity. It rather reveals specific familial or social relations, multi-lateral obligations as well as privileges. Just as giving love and taking good care of young children belong to the essential duties of parents and thus are implied in the name of "father" or "mother," taking respectful care of his or her aged parents is implied in the name of "son" or "daughter." Thus understood, fulfilling such a moral duty as taking respectful care for one's aged parents is an essential part of being a son or a daughter. That is to say, if one fails or is not willing to take respectful care of one's parents, one is not qualified to be, and thus should not be, rightly called a son or a daughter. Of course, I could fail to be the son or daughter of my parents in my family just as I could fail to be a teacher in my school or fail to be a citizen in my country. Very clearly, this Confucian justification appeals to historically, culturally, and conventionally formed norms[26] to evaluate and to adjust human activities in our social and everyday life. However, even though this is always the case, it does not necessarily follow that it ought to be the case, although often these two are not unrelated. Therefore, we have to go deeper and find some other hidden rationale, which supports this historical and cultural convention as the moral norm in Confucian Ethics.

The second Confucian justification is based on innate human moral intuitions, which can first be found in Confucius but was greatly developed in Mencius' theory of the good beginnings of human nature. In 17:21 of the *Analects*, for example, Confucius justifies the traditional custom of "three years' mourning" on the basis of "feeling at ease" (*an*),[27] which, according to Mencius' interpretation, is the beginning or "germ" of morality.

> All men have a heart which cannot bear to see the suffering of others. ... [For example,] when men suddenly see a child about to fall into a well, they all have a feeling of alarm and distress, not to gain friendship with the child's parents, nor to seek the praise of their neighbors and friends, nor because they dislike the bad reputation. Therefore, we see that a being without the feeling of commiseration is not human, a being without the feeling of shame and dislike is not human, a being without the feeling of deference and compliance is not human, a being without a feeling of right and wrong is not human. ... Men have these four beginnings just as they have their four limbs. Having these four beginnings, but saying that they cannot develop them is to destroy themselves (*Mencius*, 2a:6).[28]

Very clearly, Mencius listed here four of such good beginnings in all human beings and claimed that they are innate and intuitive. These four good beginnings establish the grounding of human moral duties and thus distinguish human beings from non-humans; e.g., beasts. In 3A:5 of *Mencius*, we are given a concrete example that illustrates how a Confucian justifies human beings' filial obligation to provide dignified funerals to their dead parents.

> Presumably there must have been cases in ancient times of people not burying their parents. When the parents died, they were thrown in the gullies. Then one day the sons passed the place and there lay the bodies, eaten by foxes and sucked by flies. A sweat broke out on their brows, and they could not bear to look. The sweating was not put on for others to see. *It was an outward expression of their innermost heart*. They went home for baskets and spades. If it was truly right for them to bury the remains of their parents, then it must also be right for all dutiful sons and men of humanity to do likewise [my italics].

This example shows, on the one hand, that children's moral duty to their parents has a natural, inborn origin within everyone's innermost heart. On the other hand, those intuitions are only beginnings. They are not always recognized by every one of us and cannot guarantee everyone will become a moral person; i.e., a man of humanity. The very fact that they are only beginnings requires us to nurture them, to cultivate and develop them. Therefore, from a Confucian point of view, we are not simply born the children of our parents, we become the children of our parents through learning and self-cultivation.[29]

The third Confucian justification can be seen from the Confucian concept of justice or righteousness (*yi*[a]). Traditionally, Confucians defined the meaning of "*yi*[a]" from the interactive relations between my "personal self" (*wo*) and my surrounding social, historical, and natural communities (*qun*). For example, Dong Zhong Shu (c.179 – c.104 B.C.E.), the most famous Confucian scholar in the Han Dynasty, defined "*yi*[a]" as follows:

> *Yi*[a] means *yi*[b] (appropriation) to one's own person. Only once one is appropriate to his own person can this be called *yi*[a] (righteousness). Thus, the expression *yi*[a] combines the notions of "appropriateness" (*yi*[b]) and "personal self" (*wo*) in one term *yi*[a]. If we hold on to this insight, *yi*[a] as an expression refers to personal self. Thus it is said that to realize *yi*[a] in one's actions is called attaining it in oneself (*zi de*); to neglect *yi*[a] in one's actions is called self-negligence (*zi shi*).[30]

According to Dong and other Confucians during his time, *yi*[a] should be defined in term of its homophone, *yi*[b], which means "right, proper, appropriate, suitable." In both classical and modern Chinese, the word *yi*[b] refers often to one's making oneself over to become appropriate to one's surrounding environments, e.g., one's familial, social, and natural communities. It refers also to making one's surrounding environments appropriate for one's self-attainment or self-accomplishment. Therefore, Dong's interpretation of *yi*[a] in terms of *yi*[b] indicates an interplay or a dialectical interaction between *yi*[a] and *yi*[b], between the personal self and its contextual and communal environments out of which an individual person reaches her identity, realization, and accomplishment.[31] Based on this conception of *yi*[a] as justice and righteousness and as the interplay between individual self and her surrounding communities, Confucians think that fulfilling one's obligations, such as being a lovely parent and taking good care of one's young children, and/or being a filial son/daughter by taking respectful care of his/her parents when the parents are old, is simply part of the way of self-realization and of self-accomplishment. Failure to do this will be called "*bu yi*" (non-righteousness). Our natural and innermost moral feelings of "*xiu*" (shame) and "*wu*" (dislike), according to Confucians (*Mencius*, 3A:5), are simply signals of both internal and social disapproval of these non-righteous actions, and thus mark the beginning of the development of righteousness and justice.[32] On the other hand, the interplay between *yi*[a] and *yi*[b] not only asks a yielding or a sacrifice of my personal self to my environmental communities in the way of appropriation, it also affirms my uniqueness in such an appropriation. That not only includes my duties but also my privileges and rights, which are due to my specific situation in my surrounding communities. Thus understood, the Confucian concepts of social justice and righteousness are not against the idea of equality and fairness among the members of the society. It is rather an affirmation of it if we consider it within a larger social and historical context.

Some people claimed that adult children's moral duty to take respectful care of their aged parents may be seen as an unfair request for the younger generation to make sacrifices for the well-being of the older generation (Daniels, 1988, pp.4-6). But if we, as a Confucian often does, take human life as an organic and dynamic process of birth, growing, flourishing, declining and dying, then the rationale behind the Confucian concept of filial obligation will become clearer. Nothing seems more natural and fair than, having received care from our parents when we were young, reciprocating this care by taking care of our parents when they are old.[33] Therefore, the charge of unfairness and inequality of Confucian filiality can only make sense on the assumption that the individuals in our social and communal life must be seen as undifferentiated, colorless, and isolated social atoms. But for a Confucian this assumption itself is questionable and unaccepted.

The fourth and the last Confucian justification may be found in the considerations of the cultural, social, and political consequences. As we know, the family was a basic social, economic and cultural unit of the society in ancient China. It played a fundamental role in regulating and stabilizing Chinese social and political life in the past, and it still plays an important role today. Family is ideally the first school of virtue, and parents are often the first teachers of their children. The values we learn from our family life, according to Confucians, will also make possible a good society. That is to say, we first learn how to deal with other people in society from watching our parents deal with each other, with our grandparents, and with us.[34] Therefore, it is very hard to imagine that a person who is devoid of caring, or unwilling to care for, her own family members can be a good citizen who will care for other people in the society. This is why in the Confucian tradition "*xiao*" (filial piety) was understood as the "root" of humanity and morality.[35]

It should be noted here that "*xiao,*" especially when we consider each of its three meanings mentioned above, was often used to justify and support the totalitarian and oppressive structure of the traditional patrilinear family and society. It is no doubt a fact that *xiao* played a very conservative political role in the past. However, when scholars point out that there was a historical connection between the kinship of the patrilinear family and the kingship of the totalitarian state (e.g., Schwartz, 1985, pp. 67-75, Roetz , 1993), they often neglect the fact that the care/love relation within a family is more natural and more primordial, and that the care/love relation between parents and children may not necessarily include patrilinear power and oppression. In today's society, for example, old age is not always associated with totalitarian political power. In many cases, especially in the case of health care for the elderly, old people are often disadvantaged and powerless. Considering this fact, a Confucian would argue that advocating the first sense of *xiao* as a virtue and adopting it as a moral duty of adult children will not only increase the happiness and security of our aged parents in their later years, but will also make members in our society care more for each other, especially for those who are disadvantaged.

Taking care of the aged generation has always been a social problem for civilized societies. The question is therefore not whether the elderly should be taken care of, but who should take care of them? There are few doubts that one has a moral duty to take care of oneself. But if a person has lost the ability to take care of herself, due either to old age, or due disease associated with old age, who, if anyone, has a moral obligation to take care of her? If Daniels and English are right in saying that adult children do not have any more of a moral obligation to take care of their aged parents than any stranger on the street, or that such an obligation only has a voluntary basis, then most likely either the burden of care would be on the whole society or the elderly who are disadvantaged would suffer. And if letting the elderly suffer is immoral, then placing the burden of caring for the elderly on the whole society (through the government) would seem to be the only option.

However, there are at least two further questions here. First, should the society have that burden? Second, can the society or the government really provide adequate care for the elderly? If I, as a son, do not have a moral duty to take care of my parents, why should I, as a stranger, have a moral duty to take care of anyone else's parents? Is the moral duty of helping a stranger based on my voluntary free will or on my existential status as a human being? If my existential status as a fellow human being imposes on me such a moral duty, why not my existential status as the son of my parents? On the other hand, the warning signals continually coming from the government-run Medicare system, as well as the Social Security system in the United States indicate that the society may not be able to bear the burden anymore without threatening the bankruptcy of the whole government. From a Confucian point of view, at least part of the problem is caused by the deterioration of the family in our modern social life. The family, as a natural institution, should play a mediating role between individuals and society. That is to say, Confucians will deny neither the existential moral duty of the elderly to care for themselves, nor that of members in the society to care for the elderly. What a Confucian wants to suggest is the addition of the familial duty fulfilled by the adult children. All three kinds of moral duties, i.e., the individual, the social, and the familial, need to work together in order to strive towards the Confucian social ideal of "*da tong*" (the Great Harmony) where

> ... [t]he elders having a happy ending, the youths having enough businesses to do, the young children having been well nurtured, and all the old men without wives, old women without husbands, old people without children, young children without parents having been taken good care of.[36]

In light of all these Confucian "justifications," I would like to conclude that a moral obligation, such as adult children's filial obligation of taking respectful care of their aged parents, should be based on a cluster of different moral considerations such as the continuation of the historical and cultural tradition, the innermost human moral intuitions, the prudential and pragmatic calculation of consequences, the nature of human life and human society as holistic/organic and communal, etc. Each of these arguments, taken individually and separately, might not be very convincing or strong enough to support a moral obligation.

However, if we consider them together, as a Confucian often does, the whole cluster of arguments becomes much stronger. Moreover, it is also not necessarily incompatible with the argument of individual autonomy, provided that we clarify the concept of consent, as I have done in the first section of this paper. Nevertheless, in comparison with these Confucian arguments and considerations, we may say that the Daniels/English thesis of "linearly" grounding our moral obligations **only** on human intentional consent is too simple and too narrow minded and thus cannot really capture the true, complicated nature of human moral obligations.[37]

Department of Philosophy
Oklahoma State University
Stillwater, USA

NOTES

1 For example, Norman Daniels told us that "In 1983 we spent ... $217 billion or $7,700 per elderly person" (Daniels, 1988, p.5).

2 In his 'Obedience and Illusion,' Michael Slote expresses a similar idea. According to Slote, it is "difficult to believe that one has a duty to show gratitude for benefits one has not requested" in O'Neill and Ruddick (1979), p.320.

3 See Jane English, 'What Do Grown Children Owe Their Parents?' in Sommers and Sommers (1993), p.763.

4 For example, Daniels says: "Children did not ask to be brought into existence" (Daniels, p.29), and calls the traditional filial relation "not self-imposed." Because of that, "we remain without compelling foundations for filial obligations, ... " (Daniels, p.34). English, though criticizing the traditional understanding of the nature of filial relation as being "reciprocal," defines filial relation as a relation of friendship. According to her, a filial relation without a friendship, which assumes mutual consent, does not endow any moral obligation. In English's words, "The relationship between children and their parents should be one of friendship characterized by mutuality rather than one of reciprocal favors" (Sommers and Sommers, p.762), and "After a friendship ends, the duties of friendship end" (Sommers and Sommers, p.761).

5 Chenyang Li made this point in his article "Grown Children's Filial Obligation," in Timothy Shanahan and Robin Wang (1996), pp. 443-447.

6 My discussion of adult children's filial obligation will exclude the case of abusive parents. A parent's abuse on her child can be seen as a case of the parent's failure of fulfilling her parental obligation, and is condemned in most moral traditions. In many cases, parental abuse of a child or a parent's failure to fulfill her parental obligation will justifiably release the child's filial obligation.

7 This idea can be traced back to Aristotle. According to Aristotle, a moral praise or blame should be based on whether an individual moral agent behaves "voluntarily or "involuntarily." "Being voluntary," Aristotle held, means that (1) an individual is internally

motivated rather than externally compelled to act; (2) the action may not be not a result of ignorance or deceit. See Aristotle, 1110a5 – 1114b15.

8 Ironically, a similar example of a good swimmer can be also found in Daniels. However, Daniels calls it "supererogatory" rather than "obligatory" (Daniels, p.33).

9 The words "existential" and "factical" should be distinguished from those of "intentional" and "factual." I use them in Heidegger's sense, which is based on his theory of Dasein as "being-in-the-world-with-others." As for Heidegger's concepts of "existence" and "facticity," see Heidegger, (1962), pp.78-86; 235-241.

10 In his *A Theory of Justice*, John Rawls makes a careful distinction between "obligation" and "natural duty." According to Rawls, both "obligations" and "natural duties" are moral requirements. Their main distinction consists in the following three aspects: (1) obligations "arise as a result of our voluntary acts" while natural duties "apply to us without regard to our voluntary acts"; (2) "the content of obligations is always defined by an institution or practice the rules of which specify what it is that one is required to do" while natural duties "have no necessary connection with institutions or social practices; their content is not, in general, defined by the rules of these arrangements"; (3) "obligations are normally owed to definite individuals, namely, those who are cooperating together to maintain the arrangement in question" while natural duties "hold between persons irrespective of their institutional relationships; they obtain between all as equal moral persons" (Rawls, p.113; p.115). On the one hand, I agree with Rawls in saying that one moral requirement arises from voluntary acts while the other does not, although I don't want to use the word "obligation" exclusively for those moral requirements based on voluntary acts. In many cases, as we know, "obligation" and "duty" mean the same in our ordinary use of English. For example, we see this in sentences such as "Citizens have an obligation to observe the laws of their country;" or "Mentally gifted people are under an obligation to develop their capacities." Therefore, I use "moral responsibility" for those moral requirements cause by voluntary acts, "moral duty" for those which are not connected with the voluntary acts, and "moral obligation" for both. On the other hand, I don't agree with Rawls when he says that the content of duties has "no necessary connection with institutions or social practice." Maybe he thinks that all social institutions, by their nature, have a voluntary or contractarian grounds. But we know that not all institutions or social practices, e.g., the family, are based on contractarian grounding. They are naturalistic social institutions. Because of that, at least some of our moral duties arise from the status we have or roles we play in a naturalistic social institution. It should also be noticed that Norman Daniels, following Rawls, mentions the distinction between the "natural duties" and the "moral obligations" (Daniels, p.29). However, it seems to me that he then quickly claims without a justification that a parental duty to children and an adult child's duty to parents belong to the category of "moral obligation," or in my term, "moral responsibility," rather than to that of "moral duty."

11 In his *Punishment and Responsibility,* H.L.A. Hart distinguishes four senses of responsibility, which are (1) Role-Responsibility; (2) Causal-Responsibility; (3) Liability-Responsibility; and (4) Capacity-Responsibility. However, Hart's discussion of the moral sense of all the four types of responsibility and his distinction between legal responsibility and moral responsibility in his discussion indicate that the intentional and voluntary consent of individuals should be the sole moral basis of all the four types of responsibility. See Hart, (1968), pp.210-230.

12 As for English translations of the *Analects*, see Lau, D.C. (1979) or Waley, A. (1989).

[13] Roetz has a detailed discussion of the three meanings of Confucius' "*xiao*." See Roetz (1993), pp.53-66.

[14] For example, Fung summarized Confucius' "*xiao*" as " ... giving them [parents] not only physical care and nourishment, but also nourishing their wills; while should they fall into error, it consists in reproving them and in leading them back to what is right. After the death of our parents, furthermore, one aspect of it consists in offering sacrifices to them and thinking about them, so as to keep their memory fresh in our minds" (Fung, 1952, p.359).

[15] According to Arthur Waley, this third sense of "*xiao*" as taking care of the spirits of dead parents might be the primitive meaning of word "*xiao*" in ancient China. As Waley observed, in the *Shi Jing* (the *Book of Songs*) "[*xiao*] refers almost exclusively to piety towards the dead. Out of twelve instances nine can only be taken in this sense." Therefore, Confucius and the Confucians during the fourth century B.C. interpreted "*xiao*" not only as the piety towards the dead but more importantly, towards living parents shows a "general transference of interest from the dead to the living which marked the break-up of the old [*Zhou*] civilization" (Waley, 1989, "Introduction," pp.38-39).

[16] Bo Yi was the eldest son of the ruler of Gu Zhu State and lived in the end of Shang Dynasty. He was called by Confucius as a moral sage or a man of humanity. As for his story, see *Analects*, 5:23; 7:15; 18:8; 16:12.

[17] Zhou[b] was the last king of Shang Dynasty and a tyrant. He was defeated by King Wu, the son of King Wen, and was killed in the war.

[18] Xi Bo was King Wen's title before he was called king.

[19] Tai Gong served as the premier for King Wen and was called a sage by later Confucians.

[20] Unless I state otherwise, all the English translations of *Mencius* in this essay follow Lau, D.C. (1970), sometimes with minor modification.

[21] See *Tang Lu Shu Yi / Dou Song*; *Tang Lu Shu Yi / Zhi Zhi.*

[22] See *Da Qing Lu Li / Ming Li.*

[23] The Chinese Marriage Law, Section 3, Article 15. I use the translation of Chenyang Li.

[24] Daniels (1988, p.28) admitted the fact too. But he wanted to call it a special case only in Confucian Asia.

[25] See Schwartz' (1985, p.92) discussion of Confucius' theory of name.

[26] It should be noticed that Confucian moral norms are, by their nature, not prescriptive or categorical, but exemplary and pedagogical. It is not imposed by Law, but through cultivating learning. I have a detailed discussion of it in my article "On the Golden Rule – From a Confucian Perspective," *Philosophy East and West* 10 (1999), forthcoming.

[27] Also see Arthur Waley's (1989, "introduction") and Li Ze-hou's (1985, p.19) discussions of "*xiao*" in Confucius.

[28] I follow Wing-Tsit Chan's translation of this paragraph in his *A Source Book in Chinese Philosophy*, p.65.

[29] As I have discussed above, "son" or "daughter" is not only a biological concept in China. It is also a social role which is affiliated with specific duties and privileges. Therefore, learning these duties and privileges belongs to the development of a human being. Lin Yutang emphasizes the importance of this learning by saying, " ... the affection for parents and grandparents is something that stands more in need of being taught by culture. A natural man loves his children, but a cultured man loves his parents" (Sommers and Sommers, pp.752-753).

[30] See Dong Zhong Shu, 8/8b; I use Hall and Ames' translation here. See Hall and Ames (1987), p.92.

[31] My understanding of the dialectical interplay between "*yi*[a]" and "*yi*[b]" benefits from Hall and Ames' insightful interpretation. This interplay, according to Hall and Ames, can be seen in that "whereas *yi*[a] denotes appropriateness to one's own person, *yi*[b] refers to appropriateness to one's context. *Yi* is the active and contributory integrating of self with circumstances, where the self originates unique activity and construes itself on its own term in a naval and creative way. ... The character *yi*[a], on the other hand, denotes the yielding or giving up of oneself and 'appropriating' meaning from the context or circumstances" (See Hall and Ames, 1987, p.98 and pp.348-349, no.51).

[32] For example, Mencius said, "The felling of shame and dislike is the beginning of righteousness" (*Mencius*, 3A:5).

[33] Here it is nothing to do with "owing" or "paying debts," as we found from reading Jane English's (Sommers and Sommers, 1993) article. According to Confucians, life should be seen as a continual stream. My parents may be seen as my life in the past and my children my life in the future. Just like it would be ridiculous to say that my hands, in providing food to my stomach, are "paying debts" to the latter because it helped to keep the hands alive, it is misleading to talk about "owing debts" between parents and children. Therefore, the difference between English and a Confucian on filial obligation does not consist in the "owing/non-owing" relation, but in that the former understands the filial obligation as a causal relation while the latter understands it as an existential relation.

[34] There is an ancient Chinese story which is very popular among Chinese. Once upon a time, there was a family of a grandfather, a father, and a son. The father did not take a good care of the Grandpa. When the Grandpa died, the father was so stingy that he took the Grandpa's dead body out with a broken basket. When the young boy saw it, he told his father: "Dad, please don't forget to bring the basket back. It is still useful." The stingy father was very happy to hear what his little son said. Then he asked his son what he would use it for. His son answered: "I will re-use it when you die."

[35] For example, we can read in the *Analects 1:2* that "Few of those who are filial sons and respectful brothers will show disrespect to superiors, and there has never been a man who is not disrespectful to superiors and yet creates disorder. A superior man is devoted to the fundamentals (the root). When the root is firmly established, the *dao* will grow. Filial piety and brotherly respect are the root of humanity (*ren*)."

[36] *Da Tong /Li Yun*; also see *Mencius*, 1B:5

[37] The first two sections of this paper were presented at the American Philosophical Association Annual Meeting, Eastern Division, at Atlanta, on December 27-30, 1996. I would like to thank Dr. Chung-Ying Cheng for his valuable comments. The Dean Incentive Grant of 97-98 from the College of Arts and Sciences at OSU supported my further research and helped me to accomplish the whole paper. My gratitude also goes to Dr. Ruiping Fan, Dr. George Graham, Dr. Edward Lawry, Dr. Chenyang Li, Dr. Robert J. Radford, Dr. Jim Allard, and Dr. Michael Zimmelmann for their support, criticisms, and helpful suggestions.

BIBLIOGRAPHY

Aristotle: 1985, *Nichomachean Ethics*, T. Irwin (trans.), Hackett, Indianapolis, Indiana.

Confucius: 1973, *Analects*, D.C. Lau (trans.), Penguin Classics, London.

Chan, W.T.: 1963, *A Source Book in Chinese Philosophy,* Princeton University Press, Princeton.

Daniels, N.: 1988, *Am I My Parents' Keeper?* Oxford University Press, Oxford.

Fung, Y.: 1952, *A History of Chinese Philosophy*, vol.I, D. Bodde (trans), Princeton University Press, Princeton.

Fung, Y.: 1948, *A Short History of Chinese Philosophy*, D. Bodde (trans), The Free Press, New York.

Hall, D. and Ames R.: 1987, *Thinking Through Confucius*, SUNY Press, Albany.

Hart, H.L.A.: 1968, *Punishment and Responsibility,* Oxford University Press, Oxford.

Heidegger, M.: 1962, *Being and Time*, J. Macquarrie and E. Robinson (trans.), Harper and Row, New York.

Li, Z.: 1985, *Zhongguo Gudai Sixiang Shilun / Essays on the Ancient Chinese Thoughts*, People's Press, Beijing.

Mencius: 1970, *Mencius,* D.C. Lau (trans.), Penguin Classics, London.

O'Neill, O. and Ruddick, W. (ed.): 1978, *Having Children*, Oxford University Press, Oxford.

Rawls, J.: 1971, *A Theory of Justice*, Harvard University Press, Cambridge, MA.

Roetz, H.: 1993, *Confucian Ethics of the Axial Age*, SUNY Press, Albany.

Schwartz, B.: 1985, *The World of Thought in Ancient China*, Harvard University Press, Cambridge, MA.

Shanahan, T and Wang, R. (ed.): 1996, *Reason and Insight*, Wadsworth, Belmont, California.

Sommers, C. and Sommers, F. (ed.): 1993, *Vice and Virtue in Everyday Life*, Harcourt, Fort Worth, TX.

Waley, A. (trans.):1989, *The Analects of Confucius*,Vintage Books, New York.

Weber, M.: 1951, *The Religion of China*, Gerth, H. (trans.and ed.), The Free Press, New York.

RUIPING FAN

JUST HEALTH CARE, THE GOOD LIFE, AND CONFUCIANISM

I. INTRODUCTION

Fashionable theories of social justice in contemporary societies engage versions of egalitarianism, utilitarianism, redistributivism, and Rawlsianism.[1] These theories have provided views of justice in general. They have also provided accounts of health care justice in particular. This essay examines a characteristic that all these theories share: an attempt to establishing an account of justice for the structure of society in separation from particular understandings of the good life for individual persons. I identify this character as "an intended separation." Specifically, under such "an intended separation," each of these theories attempts to justify its view of justice independently of any concrete premises from particular religions, metaphysics, ideologies, or conceptions of the good life. Each contends that its requirements of justice are compatible with all conflicting but reasonable accounts of the good life. Consequently, each argues that its views of justice ought to be accepted by all reasonable individuals and communities in contemporary pluralist societies.

This essay illustrates that with such "an intended separation" the contemporary theories of justice embrace a less ambitious intellectual goal than the comprehensive moral systems that formed within the Enlightenment project. The modern Enlightenment project did not plan to provide an account of justice or rightness only for the structure of society. Instead, it attempted through sound rational argument to establish a comprehensive, canonical, and universally-applicable system of morality to regulate both societal structure and individual lives. However, the modern Enlightenment project has failed rationally to secure a comprehensive morality (MacIntyre, 1984, 1988, 1990). Rational argument cannot justify a standard morality without begging the question or involving infinite regresses (Engelhardt, 1991, 1996). The contemporary theories of justice have come to form a default strategy for the modern Enlightenment project. They no longer attempt to establish a

Ruiping Fan (ed.), Confucian Bioethics, 257–284.

full-fledged account of morality. Instead, they seek only to provide guidance for political, constitutional and general social and economic arrangements, leaving all reasonable conceptions of the good life untouched. By focusing on the rightness of societal structure alone, the contemporary theories diminish the idealistic hope of the modern Enlightenment project through using the faculty of reason to disclose both rightness and goodness in human life.

Using health care as an example, this essay argues that a separation between rightness and goodness intended by the contemporary theories of justice is not feasible. It shows that this separation may have been sincerely intended, but it cannot be carried out. Section II of this essay demonstrates how the contemporary theories of justice have attempted to offer substantive guidance regarding health care allocation independent of any particular understanding of the good life. Section III shows how the intended separation in the contemporary theories of justice represents a modest retreat from the comprehensive moral programs of the modern Enlightenment project. Section IV illustrates how the apparently "thin" concept of opportunity assumed in the Rawlsian theory of justice contains a robust distinction between normal and abnormal opportunities, and that this distinction significantly bears upon some particular understandings of the good life. These three sections lead to Section V which concludes that the separation intended by the contemporary theories of justice is illusory because they have all unavoidably integrated concrete assumptions about the good life into their accounts of justice.

This state of affairs provides a heuristic for understanding Confucian regard for social justice. Confucianism has been the most influential moral tradition in East Asian areas for thousands of years. However, since early this century, the Confucian tradition has been judged as backward feudalist ideology unable to provide appropriate modern perspectives or ideas regarding social development and progress. Radical Chinese intellectuals claimed that Confucianism should be entirely abandoned because it is based upon a particular traditional conception of the good life incompatible with modern, progressivistic understandings, visions, and values regarding the good life (Tu, 1979, pp. 257-296). In recent years, East Asian scholars have enthusiastically followed contemporary Western scholars in attempting to justify an account of justice disengaged from their traditional views of the good life. The contemporary Western theories of justice, especially Rawlsianism, have attracted wide attention. However, if these contemporary understandings cannot really be justified in isolation from particular assumptions of the good life, then it is wrong to assert that contemporary Western theories of justice are more defendable than a traditional account of morality. East Asian scholars must reassess their reasons for accepting these contemporary Western theories and reconsider their objection to Confucianism.

Sections VI and VII of this essay lay out a Confucian view of social justice and its concrete implications for health care allocation. These two sections show that Confucianism is a teleological framework for governing both social structure and individual lives. Since the Confucian account of justice is intertwined with its conception of the good life, they cannot be isolated from each other without destroying their teleological framework as a coherent whole. One cannot tease out the Confucian account of justice independently of its conception of the good life; neither can one lay out the Confucian conception of the good life in separation from its account of justice. From the Confucian perspective, a life cannot be good without being just; and a society cannot be just without being good. The Confucian teleological framework of justice in its comprehensiveness stands in contrast with contemporary Western theories of justice with their intended separation.

This essay concludes that the contemporary accounts of justice do not stand in a better position than traditional Confucianism to claim universal acceptance. A significant distinction between traditional Confucianism and the contemporary theories of justice lies in their different assertions regarding the relationship between the right and the good. Confucianism concedes that its account of justice is necessarily connected with its particular view of the good life, while each contemporary theory of justice claims an intended separation between its account of justice and any concrete conception of the good life. The matter of the fact is, however, that each contemporary theory has smuggled substantive views of the good life into its particular account of justice. Although those views of the good thus absorbed may often be attenuated and even be fragmented, they remain substantive. They are derived from particular traditions, religions, or ideologies. Consequently, it is misleading to argue that those contemporary theories of justice are not parochially related to any particular understanding of the good life. We must rethink the seemingly plausible conclusion that East Asian people must accept one of such contemporary theories to guide their social arrangements in general and health care distribution in particular.

II. THE CONTEMPORARY THEORIES OF JUST HEALTH CARE: AN INTENDED SEPARATION

What does justice require a state to do with health care? This issue has become a hot topic in recent decades. Incompatible answers have been offered, based upon different theories of justice. Various versions of egalitarianism, utilitarianism, redistributivism, and Rawlsianism are often employed to provide accounts of health care justice (see, e.g., Buchanan, 1981; Brody and Engelhardt, 1987, pp. 28-33). The strong version of egalitarianism requires that the state ensure all people receive an equal level of health care appropriate to their health care needs.[2] Moreover, all people should receive all the health care from which they would benefit (Veatch, 1980, 1991). The utilitarian theory of justice, on the other hand, requires the state to establish a health care system that, in combination with other social institutions, maximizes utility. As to the questions of whether the system should be egalitarian and whether it should adopt the market mechanism, utilitarians have to appeal to empirical investigation and utility calculation to provide answers.[3] Some redistributivist theories of justice, however, demand that only a basic minimum level of health care to be guaranteed for everyone. Beyond that minimum level, such redistributivists allow individuals capable of purchasing additional health care to do so (See, e.g., Beauchamp and Childress, 1994). Finally, the dominant interpretation of John Rawls' theory of justice regarding its application in health care requires that the state provide all-encompassing health care services to fulfill Rawls' second principle of justice, particularly the requirement of fair equality of opportunity and its priority (Daniels, 1985).

These different theories indicate incommensurable understandings of social justice in general and of health care justice in particular. However, despite their mutual incompatibility in content and requirements, they also share a remarkable theoretical feature: they each intend to provide an account of justice separated from any particular understanding of the good life. Specifically, each of these theories attempts to give substantive guidance regarding just distributions in health care without touching the rationales of particular ways of life lived by individuals and their communities. Each ventures to justify its account of justice from a goodness-neutral perspective by separating its account of justice from any concrete view of the human flourishing. Accordingly, many contemporary theorists expect that no matter how many divergent, specific understandings of the good life exist in society, all individuals will accept a substantive account of justice or rightness not based upon any particular conception of the good life.

For instance, Rawls holds that his theory of justice allows for "a diversity of doctrines and the plurality of conflicting, and indeed incommensurable, conceptions of the good affirmed by the members of existing democratic societies" (Rawls, 1985, p. 225). All members of conflicting but reasonable religious, moral, and cultural communities, from Rawls' understanding, should accept his principles of justice without concern for possible contradictions between those principles and the particular doctrines of the good life that they hold. This is because, he argues, his theory does not offer a full conception of the good life capable of indicating what is of value in human life (Rawls, 1993, p. 13). Instead, his account is rooted in a "thin" theory of the good which lays out only the general prerequisites that rational persons must want in order to carry out their life plans, whatever their plans are (1971, p. 396). Norman Daniels provides a way of extending Rawls' theory to health care allocation. He borrows the principle of fair equality of opportunity from Rawls' second principle of justice. This principle, according to Daniels and Rawls, requires that the state offer national health services to satisfy people's health needs (Daniels, 1985, pp. 42-45).

From Robert Veatch's egalitarian perspective, one should consider possible equal outcomes for all the people as the requirement of justice. Veatch argues that there are fundamental egalitarian assumptions about the moral nature of the world that everyone should reasonably accept, no matter what conception of the good life one assumes (Veatch, 1991, pp. 84-86). He holds that his egalitarian theory gains a convergence of both traditional religious and contemporary non-religious convictions. Moreover, he contends, anyone who thinks from a moral point of view must accept his egalitarian theory of justice (Veatch, 1986). Under his theory, the state should guarantee that everyone have an opportunity for a level of health equal as far as possible to the health of others (1981, p. 275).

Again, certain redistributivists argue that everyone has a right to a decent minimum level of health care. They argue that this right should be fulfilled through a state-enforced universal package of health care. For some redistributivists, this right constitutes a mid-level ethical principle which reflects people's "ordinary shared moral beliefs," although different religious, metaphysical, or ideological doctrines may continue to exist in society (Beauchamp and Childress, 1994, p. 100). Moreover, they claim that maintaining a decent minimum health care reflects the convergent outcome of all moral positions, such as egalitarianism, utilitarianism, and libertarianism (1994, p. 356).

Finally, the utilitarian view of justice argues that justice requires maximizing the good, no matter how the good is defined. Different versions of utilitarianism offer different accounts of the good or utility. For instance, the hedonist version of utilitarianism claims that utility is pleasure and the absence of pain. The theory of preference-satisfaction, as another version of utilitarianism, asserts that utility is the satisfaction of individuals' preferences, whatever preferences individuals hold. Finally, the objectivist version of utilitarianism argues that utility is objective values, such as love, knowledge and wisdom, which exist objectively regardless of whether individuals cherish them or not (Moore, 1912, Ch. 7). In short, utilitarian views of justice defend themselves as being applicable to different understandings of the good. The good can be understood as pleasure, preference-satisfaction, or objective values. As a general theory of justice, all that utilitarianism essentially requires is to maximize the good, no matter which definition is given to the good, or which way of life is held to be appropriate.[4]

In short, all these contemporary theories of justice share the characteristic of attempting to separating the right from the good. In attempting to establish such a separation, each of these theories acts to justify its universal acceptance in the large-scale societies with divergent conceptions of the good. Two questions then arise regarding these theories. First, why do they pursue such a separation? Second, can they succeed in achieving such a separation? In order to answer the first question, we must take a glance at the hopes of the modern Enlightenment project.

III. THE MODERN ENLIGHTENMENT PROJECT: A CULT OF REASON

The modern age of the West was marked by the Enlightenment project which reached its height in Europe in the 18th century. It identified morality as rationality even when the content of morality was not drawn from reason itself and attempted to secure as much universality as possible for ethical claims through rational philosophical argument. It was a cult of reason. It wanted to establish "a moral community of all persons outside of any particular religious and cultural assumptions" (Engelhardt, 1996, p. viii). The most influential representative of the Enlightenment project is Immanuel Kant. For Kant, the motto of the Enlightenment is: "have courage to use your own reason!" (1980, p. 3). It is assumed that everyone is able to discern moral truth by using one's own capacity of reason alone. An enlightened person should not appeal to any particular moral tradition or rely on any particular moral community to draw one's moral principles or rules. Rather, one should only engage in rational philosophical deliberation, investigation, and exploration in order to disclose the content of morality. Such a morality should be canonical, content-full, and universally-applicable, binding all moral agents as such. It does not only give moral guidance to individual behaviors, but also provides fundamental moral principles to govern the structure of society. Indeed, it was the hope of the Enlightenment project that such a morality be justified solely by reason, accepted universally by all persons, and applied to both individual and society.[5]

Unfortunately, strikingly different moral accounts were produced by the Enlightenment project, including a variety of deontological and utilitarian approaches. Each of these accounts appears to be rationally defensible within its own terms, but yet they remain incommensurable with each other. Worse still, their original hope of setting forth a standard morality based solely on reason turns out to be an illusion. Instead of being independent of any particular moral doctrine or community, the Enlightenment moral scholars absorbed traditional moral beliefs, ideas, and rules in constructing their moral content. For instance, as a deontologist, "Kant never doubted for a moment that the maxims which he had learned from his own virtuous parents were those which had to be vindicated by a rational test" (MacIntyre, 1984, p. 44). Without these inherited maxims, Kant's deontology would be reduced to an "empty formalism," "the science of morals to the preaching of duty for duty's sake" (Hegel, 1967, p. 90). On the side of the utilitarian achieving, although a new moral goal (the greatest amount of happiness of the greatest number of people) was set up, the goal is only a formal and general slogan. Unless the concept of happiness is specified, this utilitarian slogan is nonsensical. However, one cannot define what happiness is except by recourse to a particular conception of the good life. Thus, traditional views of the good have unavoidably been reduced to fit within new utilitarian theories. Still, different traditional perspectives of happiness have been taken by different utilitarian schools, such as happiness as the presence of pleasure and absence of pain or happiness as desire-satisfaction. It is not surprising that utilitarian views vary from one to another depending upon which understanding of happiness is presupposed. Yet a calculation of the greatest good for the greatest number cannot be made without deciding how to rank good, compare preference satisfaction (e.g., rational vs. impassioned preferences, corrected vs. uncorrected preferences) as well as deciding how to discount them over time.

Not only have these Enlightenment theories integrated into their moral accounts particular moral content from traditional doctrines, they have also inherited the theoretical structure of traditional moral frameworks. Traditional ethical theories, such as Aristotelian, Christian, and Confucian moral accounts, possess the theoretical structure of teleological accounts. A teleological account defines the *telos* of human lives. It also identifies the nature of moral acts. Modern utilitarian theories inherit one major character of this teleological structure: the rightness of morality consists in achieving a *telos*. On the other hand, modern deontological theories inherit the other feature of traditional moral accounts: the moral rightness of a human act depends upon its nature rather than its consequence. What is possibly unique about modern utilitarianism is its insistence that the prospect of the maximum utility (the maximum pleasure and absence of pain or the maximum satisfaction of preferences) provides an appropriate *telos*. In contrast, most deontologists followed Kant to identify the authority of moral rules as right making conditions grounded in the nature of practical reason.

Consequently, the Enlightenment project has divided the coherent moral framework of traditional accounts into two sorts of modern moral theories: consequentialism and deontology. What is worse, many have mistaken modern consequentialism as of one fabric with traditional teleological accounts. For instance, Rawls follows W. K. Frankena in distinguishing only two types of moral theories: teleology and deontology. He defines a teleological theory by the standard that "the good is defined independently from the right, and then the right is defined as that which maximizes the good" (Rawls, 1971, p. 24). In contrast, a deontological theory is characterized as "one that either does not specify the good independently from the right, or does not interpret the right as maximizing the good" (p. 30). However, these definitions fail to distinguish teleology from consequentialism. A traditional teleological theory is not a consequentialist theory. A teleological theory assumes a *telos* set within certain right and wrong making conditions, while a consequentialist theory attends only to states of affairs. A *telos* is not a state of affairs or a piece of preference that can be achieved by any means, as modern consequentialist theories assume. A *telos* must be defined and pursued in terms of the virtues, the human traits that are needed for pursuing the *telos.* The virtues by themselves place constraints on the acceptable means for the *telos*. Hence, the acount that Rawls offers for teleology identifies it with a limited view of consequentialism, not of traditional teleological theories. A significant feature of traditional teleological theories is the inherent connection of the good with the right through the concept of the virtues. The virtues are both constitutive of an understanding of the good and presuppose a sense of the right. A life that exercises the virtues is a life that strives for the *telos* and therefore is both right and good. Accordingly, on traditional teleological accounts, it is misleading to say either, with the consequentialist, that the good is prior to the right or, with the deontologist, that the right is prior to the good (Stout, 1988, pp. 322-323, n. 9). The good and the right are simply inter-defined by each other within traditional teleological accounts of ways of life. Hence, the rich moral content of ways of life is not covered by either modern consequentialism or deontology. Moreover, it is simply wrong-headed to classify traditional teleology under the genus of consequentialism.

In short, the Enlightenment project failed in its attempt to discover the character of a standard, content-full, and universally-applicable morality in terms that did not depend on any traditional moral premise or perspective. On the one hand, a variety of incompatible new theories have been formed. They have all incorporated particular traditional maxims, notions, and ideas into their respective moral accounts. On the other hand, these new theories have integrated different features of traditional teleological frameworks into their respective moral structure. Consequently, if no traditional theory can be defended solely by reason as universally applicable, neither can any of these conflicting new theories. The modern Enlightenment project failed.

Indeed, one cannot through rational philosophical argument justify any particular moral theory as uniquely canonical for all humans without begging the question: one must assume what one needs to justify. Normative claims from different individuals in different moral communities engage different taken-for-granted moral criteria in framing the meaning and significance of their moral vision. They do not hold similar moral premises (e.g., fundamental moral principles and values and rankings of them) or rules of evidence (e.g., rules bearing upon the extent to which patterns in nature give evidence for morally binding natural laws). Neither do they share a common understanding of which persons or institutions are in authority to resolve moral controversies. In order to establish one particular moral theory as canonical, one must already have a particular moral standard. The difficulty is how to find such a standard. One cannot appeal to consequences without knowing how to compare consequences of different sorts. One cannot invoke a particular understanding of needs, interests, goals, and concerns without similarly begging the question of which should be guiding. Any particular content-full account requires a particular background foundation that takes a particular position regarding particular premises and rankings. But such a foundation itself needs a further foundation. This circumstance forecloses the possibility of avoiding begging the question or of arbitrarily asserting a particular point of departure, unless one has what is equivalent to a mystical veridical encounter with the truth (Engelhardt, 1996).

Moreover, traditional teleological theories (such as Aristotelianism) and faith doctrines (such as Christianity) are far from being supplanted by the new modern moral theories in contemporary Western society. Rather, as a sociological fact, new theories and old doctrines coexist. People continue to live in divergent faith and moral communities and hold incommensurable understandings of the good life. Contemporary Western society has become ever increasingly pluralistic in morality.

It is against this background of social diversity that the contemporary types of moral and political endeavors have emerged with the intention of separating the right from the good. Recognizing the intractable difficulty of the Enlightenment project of establishing a comprehensive system of morality to guide both individuals and society, these contemporary endeavors shift their focus to the basic structure of society. They attempt to disclose only an account of justice to regulate society, without addressing the divergent conceptions of the good life held by different religions, cultures, and ideologies. The separation between accounts of justice and concrete conceptions of the good life is engaged by the contemporary theories of justice in the hope that even if one is unable through reason to discover a standard, content-full conception of the good life for all humans, one can still use rational argument to justify some substantive principles of justice to regulate the political constitution and economic arrangements of human society.[6] Because such principles are meant to be independent of any particular view of the good life, it is held that they ought to be accepted and employed by all individuals, irrespective of the moral, metaphysical, and religious communities to which they belong. This becomes the core understanding of social justice in contemporary societies. People are expected to be united by an overarching, content-full account of the right, although they are distinguished by divergent views of the good. Rawls' theory of justice (1971) stands out as a magisterial representative of such an endeavor.

In order for his theory to be independent of any particular, full conception of the good life, Rawls begins with a "thin" theory of the good. He defines liberty and opportunity, wealth and income, and self-respect as prerequisites that rational persons must have to carry out their life plans. Whatever else, a rational person would prefer a wider to a narrower liberty and opportunity, and a greater rather than a smaller share of wealth and income (Rawls, 1971, pp. 395-398). Thus, he defines these goods in a purely instrumental way and argues that he has restricted them to the bare essentials. Given that every reasonable person wants these goods to live a reasonable life, Rawls holds that this "thin" theory of the good is compatible with any reasonable, full conception of the good life. According to his understanding, these goods simply provide the necessary rational motives to parties in the "original position" to establish the principles of justice (p. 396). And the principles of justice thus derived do not involve evaluating "the relative merits of different conceptions of the good" (p.94).

If it is true that Rawls' theory of the good is so "thin" that it can be neutrally integrated into all reasonable, full conceptions of the good life, then any other theory of justice based on a more "thick" conception of the good ought to subordinate itself to Rawls' account of justice in contemporary pluralist society, because it must be more parochial than Rawls' account. If a theory of justice is elaboration on a particular, "thick" understanding of the good life, it will not be compatible with other full conceptions of the good life. And it will not be able to require universal acceptance as Rawls' theory. It seems that Rawls' theory can be neutrally justified, while other theories can only be parochially defended within particular moral communities. Thus, all parochially defended theories of justice should conform to Rawls' neutrally justified theory. In this circumstance, it seems that any theory of justice depended on a full conception of the good life, such as the Confucian account of social justice, should not be seriously considered and employed to regulate contemporary society.

In the following I will use Rawls' theory as an example to see if the intended separation in contemporary theories succeeds. I will not try to offer a comprehensive assessment of Rawls' "thin" theory of the good. Instead, I will only examine his concept of opportunity and the related principle of fair equality of opportunity as applied by Daniels regarding health care allocation. This examination will illustrate how Rawls' and Daniels' concept of opportunity is already connected with particular understandings of the good, so that their theory of health care justice is not based on a very "thin" theory of the good as Rawls believes.

IV. THE RAWLSIAN THEORY OF JUST HEALTH CARE: A PARTICULAR SENSE OF OPPORTUNITY

Rawls offers the following two principles of justice to regulate the basic structure of society:

> First Principle: Each person is to have an equal right to the most extensive total system of equal basic liberties compatible with a similar system of liberty for all.
>
> Second Principle: Social and economic inequalities are to be arranged so that they are both: (a) to the greatest benefit of the least advantaged, consistent with the just savings principle, and (b) attached to offices and positions open to all under conditions of fair equality of opportunity (1971, p. 302).[7]

Although Rawls in his *A Theory of Justice* (1971) leaves health care unaddressed, Daniels interpreted that Rawls was to assign health care distribution to the domain of the requirement of "fair equality of opportunity" within Rawls' second principle of justice. Daniels developed this requirement into a basic principle of justice regarding health care allocation (1985, Chs. 1-3).[8] In Rawls' later work (1993), he has come to hold that, within the context of his contractarian theory, the establishment of national health care can succeed

> at the legislative stage when the prevalence and kinds of these misfortunes are known and the costs of treating them can be ascertained and balanced along with total government expenditure. The

aim is to restore people by health care so that once again they are fully cooperating members of society (1993, p.184).

Here Rawls makes it clear that health care should be part of a government's total expenditure. In other words, government has the right and obligation to collect resources to provide universal health care coverage. Now Rawls holds that his theory of justice supports a system of national health care aimed at correcting variations in citizens' physical capacities and skills, including the effects of illness and accidents on natural abilities. Such variations, for him, leave citizens with less than the minimum essential capacities required to be normal cooperating members of society. Accordingly, "fair equality of opportunity" requires the state to offer national health care to improve and restore the capacities of these citizens. Moreover, in a footnote in his *Political Liberalism* (1993), Rawls expresses his endorsement on Daniels' way of applying his theory in health care distribution:

> Here I should follow the general idea of the much further worked out view of N. Daniels in his "Health Care Needs and Distributive Justice," *Philosophy and Public Affairs*, 10 (Spring, 1981), and more completely in his *Just Health Care* (Cambridge, Cambridge University Press, 1985), Chs. 1-3 (p. 184).

Daniels' view is best summarized in the following paragraph:

> ... impairment of normal functioning through disease and disability restricts an individual's opportunity *relative to that portion of the normal range his skills and talents would have made to him were he healthy*. If an individual's fair share of the normal range is the array of his plans he may reasonably choose, given his talents and skills, then disease and disability shrinks his share from what is fair (Daniels, 1985, pp. 33-34, italics original).

Daniels argues that since disease and disability impair an individual's normal functioning and therefore restrict the normal opportunity range otherwise available to that individual, they decrease the individual's *fair* share of normal opportunity range. Accordingly, for both Rawls and Daniels, the state ought to provide national health care to maintain fairness for its members by preventing and treating their diseases and disabilities.

Evidently, Daniels' argument depends upon two assumptions. First, his concept of the normal opportunity range is based upon his concept of "normal functioning." He assumes that he is able to discover an objective account of human biological functioning without involving any value judgement.[9] Another crucial assumption implicit in his argument is that it is *unfair* for an individual to experience a limitation of the normal opportunity range due to the restrictions of disease or disability. It is this latter assumption that places the state under the obligation to maintain justice by establishing an all-encompassing health care system. However, even if we agree that disease and disability diminish the normal opportunity range, the question remains why it is unfair to experience less than the normal opportunity range because of the impact of disease, or disability?

It is easy to understand that it is unfair by a malicious or negligent act to cause some one to suffer a disease or disability. Social justice must require the state to maintain justice by punishing the injurer and forcing him to compensate the sufferer. However, it is difficult to see why the condition of experiencing less than the normal opportunity range due to disease or disability is by itself unfair. In the first place, diseases and disabilities are often caused by unexpected natural events. They do not involve any intentional human intervention. In such circumstances, it is nonsensical to say that the less opportunity range thus caused is unfair, just as it is nonsensical to say that it is unfair for hurricane to destroy one's property. Secondly, diseases and disabilities can also occur when individuals voluntarily engage in a variety of acts which are risky to their health, such as smoking, drinking, overeating, etc. In these situations, it is odd to argue that the less opportunity range caused by these acts is unfair. Finally, even if third parties have acted unfairly against someone, it is unclear why society is responsible for restoring a false circumstance. In short, even if we grant that disease and disability diminish the normal opportunity range, it remains problematic to claim that it is unfair to experience such diminishing because of disease or disability.

However, is it true that disease and disability diminish the *normal* opportunity range? This depends upon which opportunities are identified as normal. It is clearly not the case that one's disease or disability, in every natural and social situation, decreases the range of one's opportunities. To the contrary, a number of examples show that disease and disability may increase the range of one's opportunities. For instance, sickle cell tract increases the woman's chance of having viable children in an epidemic of malaria. Individuals with disability and disease tend to be exempted from services in army and thus they have more chances to save their lives in the time of war. The infertility of a prostitute can facilitate her to be more successful in her trade. Throughout history, there were often individuals who elaborately crippled themselves in order to attract others' compassion so as to become more successful beggars (Qu, 1991). There were even individuals who voluntarily accepted castration in order to procure serving positions in the Imperial palaces (Wong, *et al.*, 1973, pp. 232-234). All these diseases and disabilities, in fact, increased rather than decreased the life chances and opportunities of the sufferers.

Rawls and Daniels must show that these opportunities are abnormal: that they are abnormal either because they are helpful only for abnormal life goals, or because they are useful only in abnormal natural or social environments. To be a prostitute, a beggar, or an eunuch, they would argue, are abnormal life goals. A severe epidemic of malaria is also an unusual natural or social situation. Finally, one should not try to avoid serving the state by joining an army when this is forcibly required. Hence, although disease and disability increase certain opportunities, they are not normal opportunities. Rawls and Daniels would conclude that as long as the normal opportunity range is concerned, disease and disability decrease it.

This contention indicates the special sense of normality that Rawls and Daniels have assumed as part of their concept of opportunity. It is not that every opportunity counts. It is that only some "normal" opportunities count. It is in the sense of "normal" opportunity, rather than any opportunity, that Rawls and Daniels argue that disease and disability diminish the scope of an individual's opportunity. Accordingly, the Rawlsian principle of fair equality of opportunity relies on a particular understanding of opportunity. The significance of the principle does not only consist in its requirement that "those with similar abilities and skills should have similar life chances" (Rawls, 1971, p. 73). It also lies in a robust distinction between normal and abnormal opportunities implicit in their concept of opportunity. Without this robust assumption of normality, Rawls and Daniels could not argue that disease and disability diminish the normal opportunity range. With this assumption, however, their concept of opportunity is no longer a "thin" concept.

I will not address whether particular opportunities occasioned by disease and disability are normal or desirable. To make such judgments, one must have substantive standards. Without substantive standards, one cannot know whether the life of begging or prostituting is good or bad. Such standards are inevitably embedded in particular understandings of the good life. When Rawls and Daniels exclude life chances and opportunities generated by disease or disability as abnormal, this very exclusion indicates the "thickness" of their concept of opportunity. Thus, their concept is only compatible with some full conceptions of the good life, but not with others. Consequently, it is illusory for Rawls and Daniels to claim that their view of health care justice depends only on a "thin" concept of the good (opportunity) and thus is independent of any full understanding of the good life. Their presumed "thin" concept of opportunity turns out to be a "thick" concept.

V. RETHINKING JUSTICE: THE CONNECTION OF THE RIGHT AND THE GOOD

Rawls knows that his theory of justice fails to be compatible with all concrete views of the good life. That is why he emphasizes that a crucial factor in his political liberalism "is not the fact of pluralism as such, but of reasonable pluralism" (1993, p. 144). Specially, he asserts that his theory is compatible with all *reasonable* religions, doctrines, and ideologies. This means that any conception of the good life incompatible with Rawls' assumption is damned as unreasonable. However, what is reasonable is assumed by Rawls rather then justified. As the example of his concept of opportunity shows, his "thin" theory of the good assumes certain particular conceptions of the good life as normalities. Consequently, Rawls' theory fails to separate the good from the right as he intends.

The failure of Rawls' theory is heuristic. A similar problem exists for the other major contemporary accounts of justice. The egalitarian requirement of maintaining possible equal outcomes for all the people in the world is not neutral. It is in tension with a good capitalist's view that the good life must include enjoying one's liberty of action and assuming responsibility for one's action. On the other hand, if one is a Christian monk, one's good life should not focus on the enjoyments of this world as on God's mandates. Then one cannot be happy with the utilitarian understanding of the right as maximizing pleasure or preference-satisfaction. Finally, if one is a libertarian who recognizes integral to the good life is not being coerced, then the redistributivist view that the state has moral authority to enforce health care programs through compulsory taxation is unacceptable.

All this suggests that the right and the good may not be so easily isolated from each other as the contemporary theorists of justice have wanted to hold. The dream of justifying a particular theory of justice independent of any concrete view of the good life has turned out to be illusory, just as the original Enlightenment project that attempted to disclose one canonical comprehensive moral system through sound rational argument had to fail. We cannot shape a content-full moral theory without assuming particular premises, neither can we fashion a substantive theory of justice without touching upon concrete views of the good life. Consequently, the contemporary theories of justice, regardless of their intended separation, no longer hold a privileged epistemic position. Given their close connection with particular views of the good life, they cannot claim universal acceptance by all people. Accordingly, the traditional accounts and the contemporary theories stand in parallel regarding their respective relation to rational justification.

This result provides profound lessons for East Asian scholars. Since early this century, most East Asian scholars have judged Confucianism to be backward in its content and indefensible through rational argument. Although Confucianism has had a considerable history of influence, radical Chinese scholars wanted it to lose its intellectual and cultural strength as entirely as possible. After a series of social disasters caused by the coercive application of Marxist egalitarianism in China, the Chinese intellectuals have now begun to look for new types of Western theories of justice. The contemporary theories of justice, with their intended separation of justice from a content-full way of life, have attracted their attention. They have wanted to find a substantive theory of justice that can be universally accepted by the entire society.

If my previous argument is correct, that is, if it is illusory to attempt to establish an account of justice independent of any content-full conception of the good life, then East Asian scholars cannot reasonably expect to find one particular Western theory of justice compatible with all positions of the good life. They need instead to reconsider their rejection of Confucianism. As Confucian moral teaching is still at home in the ordinary life of East Asian people, it possesses immense moral and intellectual resources for pursuing social justice. It is an obligation of East Asian scholars carefully to explore Confucian views on social justice in general and health care justice in particular.

VI. CONFUCIANISM: A TELEOLOGICAL FRAMEWORK

Confucianism does not hold a "thin" theory of the good. Instead, it provides a comprehensive system of morality, including a full account of the good life that offers individuals detailed guidance regarding what moral goals to pursue, how to make moral decisions, and how to treat other people. It embodies a teleological structure, identifying a *telos* of the good life: "*zhi-shan*," namely, "the highest excellence" (*The Great Learning*, the text).[10] It lays out fundamental principles and virtues to direct individuals to pursue the *telos*. From the Confucian perspective, *zhi-shan* is not only a state of being perfect and acting with excellence for the individual; it is also the necessary moral basis for a peaceful and happy condition of all humans under heaven.

Confucians believe that the only way to pursue the end of *zhi-shan* is the self-cultivation (*xiu-sheng*) of individuals:

> Individuals being cultivated, their families can be well regulated. Their families being well regulated, their states can be rightly governed. Their states being rightly governed, the whole world can be made peaceable. [Hence], from the king of a state to the mass of the people, all must consider the cultivation of the individual at the root of everything (*The Great Learning*, the text).

To perform self-cultivation is to learn and exercise the virtues (*de*). For the Confucians, the virtues are the essential traits, qualities, and characteristics of humans *qua* humans. It is only through learning and practicing the virtues that individuals are able to seek *zhi-shan*, their *telos*. Possessing the virtues enables individuals to do the right thing at the right time in the right way. Accordingly, the Confucian system of morality includes an account of the virtues. It primarily concerns about the ways in which humans learn and practice the virtues.

Confucianism establishes one cardinal principle (or complete virtue) to direct the self-cultivation of human individuals. This is the principle of *ren*.[11] *Ren* constitutes the complete virtue of human individuals and sets forth the foundational principle of human society. It embodies the moral nature of humans and underlies all other possible human virtues.[12] According to Confucius (551-479 B.C.), the most important figure of Confucianism, "if one's will is set on *ren*, one will be free from wickedness" (*Analects*, 4: 4). When one has fully realized *ren*, one becomes a Confucian sage (*Analects*, 6: 28; 7: 33). From the perspective of Confucian metaphysics, *ren* is the root of human nature which Heaven (*tian*) endows into the human mind;[13] thus, every mature human mind is able to hear the call of *ren*. As Mencius (327-289 B.C.) claims, if humans were not made aware of *ren*, they would have no sense of rightness or justice. They would live their lives almost like the beasts, even if they are well fed, warmly clad, and comfortably lodged (*Mencius*, 3A: 4: 8). In short, according to Confucianism, *ren* is the embodiment of human nature in its perfect state. A human of *ren* is a perfect human.

In concrete terms, *ren* requires loving humans (*Analects*, 12: 22). In order to pursue *zhi-shan*, the human *telos*, Confucianism requires every individual to cultivate oneself by practicing the virtue of *ren*, loving all humans. From the Confucian view, each human deserves such love because each human, *qua* human, possesses the mind of *ren* in knowing, learning and practicing such love.[14] Moreover, the Confucian principle of *ren* requires one to practice love with a clear order, distinction, and differentiation. For Confucianism, different human relations convey different moral significance in relation to the principle of *ren*. One must begin with the application of *ren* in one's family, and then extend it to other social relations. This is why Confucianism emphasizes the five basic human relations. It is the Confucian belief that, for particular familial and social relations, the principle of *ren* entails further specific requirements, among which the most important are affection (*qin*) between parents and children, righteousness (*yi*[a]) between sovereign and subjects, function (*bie*) between husband and wife, order (*xu*) between older and younger, and fidelity (*xin*[b]) between friend and friend (*Mencius*, 3A: 4: 8). These relations, according to Confucianism, are the natural relations established by Heaven for the human beings. Each side of a relation should exercise the relevant particular virtue to maintain the moral nature of the relation. For instance, children should be filial (*xiao*) to their parents, while parents should be kind (*ci*) to their children. In short, not only does Confucianism require one to love everyone, but it also requires one to love with gradation and relativity of importance.

One can certainly tease out a Confucian account of social justice from the comprehensive teleological framework of the Confucian morality. Following Rawls, one can grant that justice is the first virtue of a society, while "a society is a more or less self-sufficient association of persons who in their relations to one another recognize certain rules of conduct as binding and who for the most part act in according with them" (Rawls, 1971, p. 4). A theory of social justice provides basic principles to guide political constitution and principal social and economic arrangements. For Rawls, such principles should inform the way of distributing rights and duties and determining the division of advantages from social cooperation (p. 7). But for Confucians, such principles should first and foremost offer general guidance for practicing the virtues. Through the mediation of the concept of the virtues, the Confucian concept of justice (or rightness) becomes an internal component of the Confucian conception of the good life. The virtues are not only definitive of the good life in an essential manner, but they are also implicitly representative of justice.

The Confucian principle of *ren* is the fundamental Confucian principle of social justice. It identifies the Confucian views of the rights and obligations of individuals in the family, in local community, and in society. But it does not understand these rights and duties as instrumental goods (so that they can be distributed) as does Rawls. It also illuminates the Confucian position on economic issues. The crucial point is that such a Confucian theory of justice is intertwined with its conception of the good life, reflecting a particular teleological structure. Cultivating oneself, regulating one's family, governing one's state, and making the whole world peaceable constitute the conditions of Confucian social justice as well as the particular aspects of Confucian lives. A Confucian theory of justice cannot be established or justified except based on a particular conception of the good life.

VII. THE CONFUCIAN VIEW OF JUST HEALTH CARE: SEEKING THE BEST APPLICATION OF *REN*

How does Confucianism understand health care justice? How does Confucianism distinguish normal and abnormal opportunities? The Confucian concept of opportunity must be understood embedded in the context of the Confucian good life. Given the content-full Confucian account of justice and the good life, there are certainly a range of opportunities that must be considered as unjust or immoral for individuals to pursue. In general, Confucians ought not to acquire any opportunity in opposition to the requirements of *ren* or *yi*[a] (unrighteousness). For instance, Confucius himself claimed that "riches and honors acquired by unrighteousness are to me as a floating cloud" (*Analects*, 7: 15). However, in order to offer an account of just health care, Confucians do not have to explore the empirical issue of whether disease or disability always diminishes the normal opportunity range. This may or may not be the case. But even if this is the case, it does not follow that it is unfair for one to have less than the normal opportunity range because of disease or disability.

Unlike Rawls and Daniels who fail to distinguish the unfair from the unfortunate,[15] Confucians understand that there are enormous amounts of natural and social outcomes that are very unfortunate, but not necessarily unfair. Confucians are well aware of the fact that there are destitute people. For instance, Confucians generally consider widowers, widows, solitaries, and orphans as tragic individuals (*Mencius*, 1B: 5: 3). They see the lives of those who are deaf, dumb, blind, lame, mutilated, or stunted as extremely unfortunate. They also understand that tragic events often happen to good persons. For instance, Yan Hui, Confucius' best student who "loved to learn," died of disease when he was young (*Analects*, 6: 2; 9: 6). Ran Bo Niu, another student of Confucius and a good man, suffered from leprosy (*Analects*, 6: 8). When Yan Hui died, Confucius mourned: "Heaven is destroying me! Heaven is destroying me!" (*Anaects*, 9: 8). When he visited the sick Ran Bo Niu, he sighed in sorrow: "It is the appointment of Heaven! That such a man should have such a sickness!" (*Analects*, 6: 8). Confucians see the final cause of life and death to be in the hand of the transcendent, or Heaven. They believe that "life and death have their determined appointment" (*Analects*, 12: 5). The meaning of such an appointment by Heaven usually goes beyond the grasp of the finite knowledge of the human beings. Accordingly, Confucians "do not complain [of their suffering] against Heaven" (*Analects*, 14: 37). In short, even when disease or disability destroys one's opportunities that would otherwise be available were one healthy, Confucians do not think it unfair.

This is not to say that Confucians do not advocate mutual assistance or individual charity in dealing with disease and disability. To the contrary, the requirement of loving all humans by the Confucian principle of *ren* invokes everyone's sympathetic responses to others' misfortunes and disasters. Confucius' visit to his student Ran Bo Niu in spite of Ran Bo Niu's infectious leprosy is a classical touching story. As a Confucian, one must take care of one's families, sustain one's neighbors, and help one's friends. One must have sympathy to all humans under heaven and attempt to assist them in their misfortune.

However, a crucial issue here is whether the principle of *ren* requires governments to establish a single-tier national health care system to address the medical needs of individuals and their families. For Confucians, the answer to this question must depend upon whether the establishment of such a system is the best application of the principle of *ren* to health care. As we discussed in the last section, the principle of *ren* requires one to apply one's love to all humans, but with distinction, order, and relativity of importance. From the Confucian understanding, the application of one's love begins with one's family in the context of a local Confucian community. According to the Confucian ideal, the best community is a well-field district[16] in which people learn the virtue of *ren* by taking care of each other:

> In the field of a district, those who belong to the same nine squares render all friendly offices to one another in their going out and coming in, aid one another in keeping watch and ward, and *sustain one another in sickness*. Thus the people are brought to live in affection and harmony (*Mencius*, 3A: 3: 18, italics added).

This Confucian thought emphasizes that people in the local district voluntarily sustain each other in sickness. It does not suggest that government may or should coercively collect resources through heavy taxes to ensure equal and universal health care for everyone. From the Confucian teaching, just as individuals should always act according to *ren* and become humans of *ren*, government should always rule according to *ren* and become a government of *ren*. Thus, Confucians do not support any government policy that extorts money from people. As Mencius states, "a government of *ren* ... must make the taxes and levies light" (*Mencius*, 1A: 5: 3). To be sure, government needs to maintain social order. It must protect people from stealing, robbery, fraud, battery, murder, and the like. But it would better not take positive actions beyond this. It should not enforce a government-controlled economy or a government-planned system of redistribution. For Confucius, a ruler who rules without taking any positive actions (*wu-wei-er-zhi*) is the ideal ruler. He admired the legendary ancient kings and sages, Yao and Shun, as the ideal rulers: "Shun was instanced having governed efficiently without taking any positive actions" (*Analects*, 15: 3).

In summary, Confucians cannot support the establishment of a government-controlled, single-tiered, and all-encompassing national health care system. Such is in conflict with the principle of *ren*. First, it is wrong for a government to enforce such a system in the name of justice or fairness, because suffering disease or sickness, although unfortunate, is not necessarily unfair. Second, the principle of *ren* requires the government to rule without taking positive actions. Government is not justified in extorting people's resources for coercive redistribution. Finally, an ideal Confucian life is primarily lived with family members. It is also lived with members of a local community (rather than a large-scale society), in which people apply the requirement of love under the principle of *ren* by taking care of each other, including providing mutual sustenance in sickness. All of this convincingly suggests that Confucianism holds a family-centered and community-oriented view of just health care. People should be left in their local community, with their own resources, freely and cooperatively to pursue an appropriate pattern of health care for themselves.

In fact, state-imposed accounts of justice have caused tremendous problems. Chinese people have witnessed how the Communist egalitarian ideal led to the state-controlled, centrally-planned, and government-run Chinese health care system with its disastrous consequences. Since the 1950s, the Chinese government has attempted to establish an equal and universal health care system through aggressive government undertakings and political movements. It attempted to ensure egalitarian health care by exempting people's responsibilities and depriving their free choices. In rural areas, the government appointed "barefoot doctors" and imposed so-called "cooperative medical service." It also forced urban health care professionals to move to the countryside in order to eradicate health care differences between cities and villages. In urban areas, the government used public resources to establish a state medical insurance for employees of state-owned companies. This insurance did not allow subscribers the freedom to choose among hospitals or physicians. Due to the lack of resources, the low quality of care, and corruption, the cooperative medical service in the rural areas entirely collapsed in the early 1980's (Hillier, 1983). About 800,000,000 Chinese peasants were left without any health insurance. Moreover, ever increasing inefficiency, waste, and corruption has been involved in the state medical insurance. But it currently privileges less than twenty percent of the Chinese population with free health care through public resources. Confucius would lament this poor health care situation in China.

VIII. CONCLUDING REMARKS: CONFUCIANISM AND BEYOND

This essay lends theoretical credence to Confucianism. The Confucian account of justice is not uniquely parochial because of its close connection with a particular conception of the good life. The truth of the matter is that the so-called neutrally applicable theories of justice, such as Rawls' theory, are not neutral at all. They are also related to specific understandings of the good life. Even if the understandings of the good life on which they rely are not as systematic as the Confucian conception of the good life, they are still not neutral and not compatible with all other conceptions of the good life. The particular ideas of the good life they incorporates are often hidden and even fragmented. Certainly this should not establish them as more reasonable or as having more universal claims.

Finally, as Confucianism is still influential in countries like China, it is theoretically legitimate and practically fruitful for people to revive its cultural and moral force and reshape a solid Confucian community. Since China is in transition to a free society, Confucian ideals can fill the vacuum of values left by the fading of the currently enforced modern Western ideologies. The deeply rooted Chinese Confucian tradition must bring its intellectual and cultural resources into full play to facilitate this transition. Regarding the issue of health care allocation, although Confucianism is a comprehensive moral system in which an account of justice and a particular conception of the good life are intertwined, we can still clearly trace its implications for the system of health care that a state should adopt.

Center for Medical Ethics and Health Policy
Baylor College of Medicine
Houston, Texas, USA

NOTES

[1] This paper addresses a broad sense of social justice, i.e., an ethically right society on balance, which takes all important considerations into account. Hence the terms "rightness" and "justice" are used interchangeably in the text. For a distinction between the broad sense and narrow sense of social justice, see W. K. Frankena (1962, pp. 1-3).

[2] Compared to Veatch's egalitarianism, Rawls' theory of justice can be considered a weak version of egalitarianism. According to his second principle of justice, inequalities can (and should) be arranged for the benefit of the least advantaged members of society, as long as the requirements of fair equality of opportunity and just savings are satisfied.

[3] Utilitarians have offered sharply contrasting accounts depending on their different ways of utility calculation. For instance, Peter Singer wants an absolutely egalitarian system of health care distribution (1976), while David Friedman wants a totally free market pattern of distribution (1991).

[4] For instance, Peter Singer argues that everyone should take "the point of view of the universe," according to which the interests of all sentient beings should be equally considered. Since sufferings and pains are the most obvious disvalues that damage interests, Singer contends that everyone is morally required to join in a cause to reduce sufferings and pains (1995, pp. 222-232). He certainly believes that this utilitarian view is compatible with a great number of different moral causes and life styles, as long as they intend or contribute to reduce sufferings.

[5] Enlightenment thinkers sought to disclose universal principles to govern both social structure and individual lives. The good examples in this regard are the accounts of John Locke and Immanuel Kant. For Locke, from the presumed intention of the Creator it followed that men were naturally equal, in the sense that no one had more power or jurisdiction than another, and were naturally free "to order their actions, and dispose of their possessions and persons, as they think fit, within the bounds of the law of nature," which forbids anyone harming another or destroying himself, and requires each to try "when his own preservation comes not in competition" to preserve the rest of mankind (Locke, 1980, §§4-6). Hence the law of nature constitutes the fundamental moral principle regulating individual lives and guiding the establishment of government and its undertakings. Kant also belongs to the natural-law tradition. He argues that there is the fundamental moral law that can be revealed by reason. Both a theory of justice for society and a theory of virtue for individuals can be derived from this moral law (Kant, 1965).

[6] The contemporary liberal principle of neutrality (i.e., the principle that requires that political theories and decisions employed in pluralist society should be neutral to any particular conception of the good life) can be placed within this genre of intended separation. For a brilliant examination of the principle of neutrality, see George Sher (1997, Ch. 2).

[7] In his later work (1993, p. 291) Rawls gives more accurate expression to his two principles of justice. For the sake of simplicity, I omit addressing any issues relating to the different expressions of his principles of justice. The different expressions do not influence the argument in this essay.

[8] Daniels proposes four different ways in which to apply Rawls' theory of justice to health care: (1) fit health care needs into the consideration of the index of primary social goods; (2) treat health care as a primary social good; (3) use the fair equality of opportunity clause of Rawls' second principle and its priority; and (4) leave health care services as something to be purchased by one's fair share of social products (1979, pp. 182-183). He himself chooses strategy (3). In addition, Ronald Green places considerations of health care under Rawls' liberty principle. See Green (1976; 1983).

[9] Rawls' and Daniels' assumption presupposes that they can offer a universally applicable notion of disease independent of any particular value judgments. Daniels argues in this line in his (1985). However, it is illusory to believe that there is such a universally justifiable conception of disease without involving any specific values. I simply omit addressing this issue in the text. For a profound exploration of this issue, see Engelhardt (1996, Ch. 5).

[10] In its formal structure, Confucianism is similar to Aristotelianism. Both contain a teleological framework in their accounts. A teleological framework holds that humans have an end (*telos*) in their lives, which is the well-functioning of themselves as humans. Both accounts of

justice and the good life must be made around such a *telos*. For a comparative study and exploration of the differences between Confucianism and Aristotelianism, see MacIntyre (1991).

[11] *Ren* has been translated into many different English terms, such as benevolence, love, altruism, kindness, charity, compassion, magnanimity, perfect virtue, goodness, true manhood, manhood at it best, human-heartedness, humaneness, humanity, etc. These multiple distinct translations reflect that *ren* is an exceedingly complicated concept. I will leave this crucial concept untranslated in the text. To my knowledge, in the extensive literature dealing with the meaning of *ren* and its evolution, an essay by Wing-tsit Chan (1955) provides the most clear and cogent account. This essay generally follows Chan's interpretation of *ren*.

[12] For instance, Confucius claims: "men of *ren* are sure to be brave, but those who are brave may not always be men of *ren*" (*Analects*, 14: 5).

[13] There has been scholarly disagreement about the Confucian view of *tian*. Some show that *tian* is a personal deity, others contend that *tian* is a transcendent force, and still others argue that *tian* is no more than the natural order. However, focusing on the texts by Confucius and Mencius, it is clear that *tian* has personal characteristics, even if it is not a person (e.g., *tian* has will), *tian* has an ethical dimension (e.g., *tian* is the source of *ren* and *yi*[a]), and *tian* is the final determination of something beyond human control (such as life, death, wealth, and honor). For a most recent study of Confucian *tian*, see Shun (1997, pp. 208-210). For a comparative study between Chinese understandings of *tian* and the Judeo-Christian notion of God, see Tu Li (1978). For a general discussion of traditional Chinese metaphysics, see Thomas H. Fang (1967).

[14] For Confucians, the objects of this sort of love are only human beings. It should not be extended to non-humans. The following story of Confucius is well-known: 'The stable being burned down when he was at court, on his return he said, "has any human been hurt?" He did not ask about the horses' (*Analects*, 10: 12).

[15] For a very helpful explication of the distinction between the unfortunate and the unfair, see Engelhardt (1996, pp. 382-384).

[16] The well-field system is the Confucian ideal of basic social and economic structure. This system might have been adopted in the former Zhou dynasty (c. 11-8th cen., B.C.). According to Mencius, the system was designed like this: "A square *li* covers nine squares of land, which nine squares contain nine hundred *mu*. The central square is the public field, and eight families, each having its private hundred *mu*, cultivate in common the public field. And not till the public work is finished, may they presume to attend to their private affairs (*Mencius*, 1A: 3: 19). For an excellent discussion of the modern Chinese debate concerning the well-field system, see Levenson, 1960.

REFERENCES

Aristotle: 1985, *Nicomachean Ethics*, Terence Irwin (trans.), Hackett Publishing Company, Indianapolis.

Beauchamp, T. L. and Childress, J. F.: 1994, *Principles of Biomedical Ethics*, Oxford University Press, New York.

Brody, A. B. and Engelhardt, H. T. Jr.: 1987, *Bioethics: Readings and Cases*, Prentice-Hall, Inc., Englewood Cliffs.

Buchanan, A.: 1981, "Justice: A philosophical review," in Earl E. Shelp (ed.), *Justice and Health Care*, D. Reidel Publishing Company, the Netherlands, pp. 3-21.

Chan, Wing-tsit: 1955, "The evolution of Confucian concept Jen," *Philosophy East and West*, Vol. 4 (January 1955), pp. 295-315.

Confucian Analects, The Great Learning and The Doctrine of the Mean, 1971, James Legge (trans.), Dover Publications, Inc., New York.

Daniels, N.: 1979, 'Rights to health care and distributive justice: Programmatic Worries', *The Journal of Medicine and Philosophy* 4, pp. 174-191.

Daniels, N.: 1985, *Just Health Care*, Cambridge University Press, Cambridge.

De Bary, Wm. T., et al. (eds.): 1960, *Sources of Chinese Tradition*, Vol. I, Columbia University Press, New York.

Engelhardt, H. T. Jr.: 1996, *The Foundations of Bioethics*, second edition, Oxford University Press, New York.

Fang, T. H.: 1967, 'The world and the individual in Chinese metaphysics,' in Charles A. Moore (ed.), *The Chinese Mind*, East-West Center Press, pp. 238-263.

Frankena, W. K.: 1962, 'The concept of social justice,' in R. B. Brandt (ed.), *Social Justice*, Prentice-Hall, Inc., Englewood Cliffs.

Friedman, D.: 1991, 'Should medicine as commodity? An economist's perspective,' in T. J. Bole III and W. B. Bonderson (eds.), Rights to Health Care, Kluwer Academic Publishers, Dordretcht, pp. 259-305.

Green, R.: 1976, 'Health care and justice in contract theory perspective,' in R. Veatch and R. Branson (eds.), *Ethics and Health Policy*, Ballinger, Cambridge, pp. 111-126.

Green, R.: 1983, 'The priority of health care,' *The Journal of Medicine and Philosophy* 8, pp. 373-380.

Hegel, G. W. F.: 1967, *Hegel's Philosophy of Right*, T. M. Knox (trans.), Oxford University Press, London.

Hiller, S. M.: 1983, 'The cultural revolution and after - Health care 1962 - 1982,' in S. M. Hiller and J. A. Jewell (eds.), *Health Care and Traditional Medicine in China*, 1800-1982, Routledge & Kegan Paul, London.

Kant, I.: 1965, *The Metaphysical Elements of Justice*, J. Ladd (trans.), Macmillan Publishing Company, New York.

Kant, I.: 1980, 'What is Enlightenment?' in L. W. Beck (ed.), *On History*, Bobbs-Merrill Educational Publishing, Indianapolis, pp. 3-10.

Legge, J.: 1971, "The Prolegomena," *Confucian Analects, The Great Learning and The Doctrine of the Mean*, 1971, James Legge (trans.), Dover Publications, Inc., New York, pp. 1-135.

Levenson, J. R.: 1960, 'Ill wind in the well-field: The erosion of the Confucian ground of controversy, in A. F. Wright (ed.), *The Confucian Persuasion*, Standard University Press, Standard, pp. 268-287.

Li, D.: 1978, *The Way of Heaven and God in Chinese and Western Philosophical Thought*, Lien-ching chu-ban shi-ye gong-si, Taiwan.

Locke, J.: 1980, *Second Treatise of Government*, Hackett Publishing Company, Indianapolis.

MacIntyre, A.: 1984, *After Virtue*, second edition, University of Notre Dame Press, Notre Dame.

MacIntyre, A.: 1988, *Whose Justice? Which Rationality?* University of Notre Dame Press, Notre Dame.

MacIntyre, A.: 1990, *Three Rival Versions of Moral Enquiry*, University of Notre Dame Press, Notre Dame.

MacIntyre, A.: 1991, 'Incommensurability, truth, and the conversation between Confucians and Aristotelians about virtues,' in E. Deutsch (ed.), *Culture and Modernity*, University of Hawaii Press, pp. 104-122.

Mencius: 1970, *The Works of Mencius*, James Legge (trans.), Dover Publications, Inc., New York.

Moore, G. E.: 1912, *Ethics*, Oxford University Press, London.

Qu, Y.: 1991, *Various Aspects of Chinese Beggars*, Yun Long Press, Taipei.

Rawls, J.: 1971, *A Theory of Justice*, Harvard University Press, Cambridge.

Rawls, J.: 1985, "Justice as fairness: political not metaphysical," *Philosophy and Public Affairs* 14 (Summer 1985), pp. 223-257.

Rawls, J.: 1993, *Political Liberalism*, Columbia University Press, New York.

Sher, G.: 1997, *Beyond Neutrality: Perfectionism and Politics*, Princeton University Press, forthcoming.

Shun, K.: 1997, *Mencius and Early Chinese Thought*, Stanford University Press, Stanford.

Singer, P.: 1976, 'Freedom and utilities in the distribution of health care,' in R. M. Veatch and R. Branson (eds.), *Ethics and Health Policy*, Ballinga Publishing Company, Cambridge, pp. 175-193.

Singer, P.: 1995, *How Are We to Live?* Prometheus Books, New York.

Stout, J.: 1988, *Ethics After Babel*, Beacon Press, Boston.

Tu, W.: 1979, *Humanity and Self-Cultivation: Essays in Confucian Thought*, Asian Humanities Press, Berkeley.

Veatch, R.M.: 1980, *A Theory of Medical Ethics*,

Veatch, R. M.: 1986, *The Foundations of Medical Ethics*, Oxford University Press, New York.

Veatch, R. M.: 1991, 'Justice and the right to health care: An egalitarian account,' in T. J. Bole III and W. B. Bonderson (eds.), *Rights to Health Care*, kluwer Academic Publishers, Dordrecht, pp. 83-102.

Wong, K. C. and Wu, L. T.: 1973, *History of Chinese Medicine*, second edition, AMS Press, Inc., New York.

CHINESE GLOSSARY

an 安

Bai Ju Yi 白居易
Bao Pu Zi - Nei Pian 《抱朴子。 篇》
Bao Shu Ya 鲍叔牙
ba yu jiao yang 八月骄阳
Bei Ji Qian Jin Yao Fang 《备急千金要方》
Beijing 北京
ben cao 本草
Ben Cao Gang Mu 《本草纲目》
Ben Cao Shi Yi 《本草拾遗》
bi guan 闭观
bian tong 变通
bie 别
Bo Yi 伯夷
bu yi 不义

cang 藏
chang sheng bu si 长生不死
Chen Cang Qi 陈藏器
Chen Cun 陈村
Chen Guang Lei 陈光磊
Chen Liang 陈亮
Chen Liang Ji 《陈亮集》
Chen Qian Chu Xian Sheng Mu Zhi Ming 《陈乾初先生墓誌铭》
Chen Que 陈确
Chen Tian Hua 陈天华
Cheng 诚
Cheng-Chung Shu Chu 正中书局
Cheng Hao 程颢
Cheng I 程颐
Cheng Shu De 程树德

Ruiping Fan (ed.), Confucian Bioethics, 285–297.

Cheng Ying	程婴
Cheng-Zhu	程朱
Chou Fu	丑父
Chu	楚
Chu Ci Ji Zhu	《楚辞集注》
Chuang Tzu (Zhuang Zi)	庄子
Chuang Tzu (Zhuang Zi)	《庄子》
Chun Qiu Fan Lu	《春秋繁露》
Chun Qiu Fan Lu - Xun Tian Zhi Dao	《春秋繁露。循天之导》
ci	慈
Da Kuang	《大匡》
Da Qing Lu Li - Ming Li	《大清律令。名例》
da ti	大体
da tong	大同
Da Xue	《大学》
Da Zheng Xin Xiu Da Zang Jing	《大正新修大藏经》
dao (tao)	道
dao bu yuan ren	道不远人
dao jia	道家
dao li	道理
dao xue	道学
dao yi	道义
de	德
de xing	德行
Diao Qu Yuan Fu	《弔屈原赋》
dong	动
Dong Zhong Shu	董仲舒
Du Si Shu Da Chuan Shuo	《读四书大全说》
Duan Wu	端午
Du Shu	《读书》
Er Ya	《尔雅》
fan guan	反观

Fan Li Sao 《反离骚》
feng, han, shu, shi, zao, huo 风寒暑湿燥火
Fu Lei 傅雷
fu zuo 俯坐
Fun You-lan 冯友兰

Gao Seng Zhuan 《高僧传》
Ge Hong 葛洪
ge wu, zhi zhi, cheng yi, zheng xin, xiu shen, qi jia, zhi guo, ping tian xia 格物致知诚意正心修身齐家治国平天下
Gong Ting Xian 龚廷贤
gu dai zhong guo ren de jia zhi guan: Jia zhi qu xiang de chong tu ji qi jie xiao 古代中国人的价值观：价值取向的衝突及其解消
Gu Yan Wu 顾炎武
Gu Zhu 孤竹
guan xing 观性
Guan Yu 关羽
Guan Zhong 管仲
Guan Zi 管子
Guan Zi - Nei Ye Pian 《管子。内业篇》
gui shen 鬼神
Guo Dai Dong 郭大东

Han 汉
Han Xue Yan Jiu Zhong Xin 汉学研究中心
Han Yu 韩愈
he 和
He Xian Ming 何显明
Hua Shan Wen Yi 《华山文艺》
Hua Wen Shu Ju 华文书局
Hua Zhong Li Gong Da Xue 华中理工大学
Huang Di Nei Jing Su Wen 《皇帝内径素问》
Huang Jun Jie 黄俊傑
Huang Ru Cheng 黄汝成
Huang Zhong Mo 黄中模

Huang Zi Ping	黄 子平
Huang Zong Xi	黄 宗羲
jen (ren)	仁
Jia Yi	贾 谊
Jiao Xun	焦循
jie cao	节 操
jie lie	节 烈
jing[a]	敬
jing[b]	静
jing[c]	经
jing zuo	静坐
Jiu	纠
Juan You Lu	《倦游录》
jue ming ci	绝 命辞
jun zi	君子
ke ji	克 己
Lao She	老舍
lao she zhi si	老舍之死
Lao Tzu (Lao Zi)	老子
Lao Tzu (Lao Zi)	《老子》
Lao Zi (Lao Tzu)	老子
Lao Zi (Lao Tzu)	《老子》
li[a]	理
li[b]	礼
Li Bai	李白
Li Ji	《礼记》
Li Ji - Li Yun	《礼记。礼运》
Li Shi Zhen	李时珍
li xue	礼学
li yi fen shu	理 一分殊
liang jiao yang	两 脚 羊
Liang Ju Chuan	梁巨 川
Liang Qi Chao	梁启超
liao ning jiao yu	辽宁 教育

Lin Yu-sheng 林毓生
Lin Yu-tang 林语堂
Lin Yuan Hui 林元辉
Ling Shu - Zhong Shi 《灵枢。终始》
Liu Qiu 琉球
Liu Zong Zhou 刘宗周
Lu Ming Fang 吕明方
Lu Xi Zhe 吕希哲
lun fo jiao de zi sha guan 论佛教的自杀观
Lun Yu Ji Shi 《论语集释》
Luo Ji Zu 罗继祖

mai shen mai de qian nian ming: lun zhong guo ren de zi sha yu min yu 卖身买得千年名：论中国人的自杀与名誉
Mencius (Meng Zi) 孟子
Mencius (Meng Zi) 《孟子》
Meng Zi (Mencius) 孟子
Meng Zi (Mencius) 《孟子》
Meng Zi Zheng Yi 《孟子正义》
ming 命
ming fen 名分
ming mu 瞑目

neng yang 能养
pin mu 骈拇

qi (Qi) 气
qian gu jian nan wei yi si: tan ji bu xie lao she, fu lei zhi si de xiao shuo 千古艰难唯一死：谈几部写老舍、傅雷之死的小说
Qian Jin Fang 《千金方》
qigong (Qigong) 气功
qi jie 气节
Qi Lin Chu Ban She 麒麟出版社
qi xiang 气象
qi zhi 气质
Qin 秦

qin	亲
qin qin	亲亲
Qing	清
qing[a]	情
qing[b]	顷
Qing Hua Da Xue	清华大学
qing qi	清气
qing xin	清心
quan	权
qun	群
qing sheng	轻生
Qu Yuan	屈原
Qu Yuan Wen Ti Lun Zheng Shi Gao	《屈原问题论争史稿》
ren (jen)	仁
Ren An	任安
ren de	仁德
Ren Min Chu Ban She	人民出版社
Ren Min Wen Xue	《人民文学》
Ri Zhi Lu	《日知录》
Ri Zhi Lu Ji Shi	《日知录集释》
ru	儒
ru jia	儒家
san bu xiu	三不朽
san min	三民
sha shen cheng ren	杀身成仁
sha shen qiu ren	杀身求仁
Shanghai	上海
Shanghai Ci Shu Chu Ban She	上海辞书出版社
Shanghai Shu Ju	上海书局
Shanghai Wen Hua Chu Ban She	上海文化出版社
Shanghai Wen Xue	《上海文学》
Shao Hu	召忽

Shao Yong	邵雍
she shen qu yi	舍身取义
shen[a]	身
shen[b]	神
Shen Nong Ben Cao	《神农本草》
Shen Nong Ben Cao Jing	《神农本草经》
shen bu you ji	身不由己
shen ti li xing	身体力行
shen xin jiao cui	身心交悴
sheng	圣
Shenyang	沈阳
Sheng Ji Zong Lu	《圣济总录》
shi	士
Shi Dao Xuan	释道宣
Shi Heng Qing	释恒清
Shi Hui Jiao	释慧皎
Shi Ji	《史记》
Shi Ji: San Huang Bu Ji	《史记。三皇补记》
shi jie	士傑
Shi Qi	施杞
shi yu you zheng	施於由政
Shi Yue Wen Yi Chu Ban She	十月文艺出版社
Shi Zan Ning	释赞宁
Shijiazhuang	石家庄
shou zu	手足
shu	恕
Shuo Zhong Hua Min Zu Zhi Hua Guo Piao Ling	《说中华民族之花果飘零》
Si Bu Bei Yao	《四部备要》
si: gei "wen ge"	死：给"文革"
Si Ma Zhen	司马贞
Si Shu Zhang Ju Zhu	《四书章句注》
si jian	死谏
si jie	死节
Si Jie Lun	《死节论》
Si Ma Qian	司马迁

Si Wang Zhi Si Yu Si Wang Zhi Shi
《死亡之思与死亡之诗》
si yu 私欲
Song 宋
Song Gao Seng Zhuan 《宋高僧传》
Song Jian 宋銒
Song Shi 《宋史》
Song Yuan Xue An - Heng Pu Xue An
《宋元学案。横浦学案》
Song Yuan Xue An - Heng Qu Xue An
《宋元学案。横渠学案》
Song Yuan Xue An - Hui Weng Xue An Shang
《宋元学案。晦翁学案上》
Song Yuan Xue An - Shuo Zhai Xue An
《宋元学案。说齐学案》
Song Yuan Xue An - Ying Yang Xue An
《宋元学案。荥阳学案》
Su Shu Yang 苏叔阳
Su Wen - Ba Zheng Shen Ming Lun Pian
《素问。八正神明论篇》
Su Wen - Shang Gu Tian Zhen Lun Pian
《素问。上古天真论篇》
Su Wen - Si Qi Tiao Shen Da Lun Pian
《素问。四气调神大论篇》
Su Wen - Tiao Jing Lun Pian 《素问。调经篇》
Sun Si Miao 孙思邈

Tai 泰
Tai Da Zhe Xue Lun Ping 《台大哲学论评》
Tai Gong 太公
tai ji 太极
Taipei 台北
Taiwan 台湾
Taiwan Shang Wu Yin Shu Guan
台湾商务印书馆
Tang 唐
Tang Jun Yi 唐君毅
Tang Lu Shu Yi - Dou Song 《唐律疏议。斗讼》

Tang Lu Shu Yi - Zhi Zhi	《唐律疏议。职制》
Tang Zhong You	唐仲友
tao (dao)	道
ti[a]	体
ti[b]	悌
ti cha	体察
ti hui	体会
ti ren	体认
ti ren	体仁
ti xing	体行
tian	天
tian di	天地
tian ming	天命
tian nian	天年
tian ren he yi	天人合一
tian xia	天下
Tong Jian Gang Mu	《通鉴纲目》
Tu Wei-Ming	杜维明
Wai Tai Mi Yao	《外台秘要》
Wan Li	万历
Wan Li Shu Dian	万里书店
wan wu	万物
wan wu jie bei yu wo	万物皆备於我
Wan You Wen Ku	《万有文库》
Wang Chuan Shan	王船山
Wang Fu Zhi	王夫之
Wang Guo Wei	王国维
Wang Guo Wei Zhi Si	《王国维之死》
Wang Jing An Xian Sheng Mu Qian Dao Ci	《王静安先生墓前悼词》
Wang Shen Pian	《亡身篇》
Wang Tao	王焘
Wang Wen Cheng Gong Quan Shu	《王文成公全书》
Wang Yang Ming	王阳明
Wang Zeng Qi	汪曾祺

wen	文
Wen Tian Xiang	文天祥
Wen Yuan Ge Si Ku Quan Shu	《文渊阁四库全书》
wo	我
wu	恶
wu chang	五常
wu gu	五谷
Wu Guang Ming	吴光明
wu jing	五经
wu lun	五伦
Wu Qiu	吴球
wu se	五色
wu sheng	五声
wu ti	五体
wu wei	五味
wu wei	无违
wu wei er zhi	无为而治
wu xing	五行
wu zhi	五志
Wuchang	武昌
xi, nu, ai, le, bei, kong, jing	喜怒哀乐悲恐惊
Xi Bo	西伯
xiang rou	想肉
xiao	孝
Xiao Bai	小白
Xiao Jing	《孝经》
xiao ren	小人
xiao ti	小体
xin[a]	心
xin[b]	信
Xin Chou Yan He Zou Za Er	《辛丑延和奏紥二》
xin xing	心性
xing cha	性察
xing qi dao	行其道
xiu	羞

xiu shen	修身
xu[a]	序
xu[b]	虚
xu lao	虚劳
xu ruo	虚弱
xu sun	虚损
Xue Gao Seng Zhuan	《续高僧传》
xue lin	学林
xun	殉
xun dao	殉道
xun fu	殉夫
xun guo	殉国
xun jie	殉节
xun jun	殉君
xun qing	殉情
xun si	殉死
xun zang	殉葬
xun zhu	殉主
Xun Zi	荀子
Yan Xin	严新
Yan Yuan	颜元
yang[a]	阳
yang[b]	养
yang kou ti	养口体
yang ling	养灵
yang xing	养性
Yang Xiong	杨雄
yang zhi	养志
Ye Shi	叶适
yi[a]	义
yi[b]	宜
Yi Shen Pian	《遗身篇》
yi shen shi fa	以身试法
yin	阴
yin mao	阴毛
Yin Wen	尹文

yin yang 阴阳
yin yang yi bing 阴阳易病
yong 用
You Yang Za Zu 《酉阳杂俎》
Yuan 元

Zeng Shen 曾参
Zeng Xi 曾皙
Zeng Yuan 曾元
Zhang Jiu Cheng 张九成
Zhang San Xi 张三夕
Zhang Zai 张载
zheng ming 正名
Zheng Xiao Jiang 郑晓江
zhi[a] 志
zhi[b] 智
zhi shan 至善
zhi xing he yi 知行合一
zhong[a] 中
zhong[b] 忠
Zhong Guo Gu Da Ming Ju Ci Dian
《中国古代名句辞典》
Zhong Guo Ren De Jia Zhi Guan Guo Ji Yan Tao Hui Lun Wen Ji
《中国人的价值观国际研讨会论文集》
Zhong Guo Ren De Si Wang Xin Tai
《中国人的死亡心态》
Zhong Guo Si Wang Zhi Hui 《中国死亡智慧》
Zhong Guo Wen Hua Yu Shi Jie
《中国文化与世界》
Zhong Guo Wen Zhe Yan Jiu Ji Kan
《中国文哲研究集刊》
zhong he 中和
Zhong Hua Shu Ju 中华书局
zhong yong 中庸
Zhong Yong Zhang Ju Zhu 《中庸章句注》
Zhou[a] 周
Zhou[b] 纣

Zhou Dun Yi	周敦颐
zhu lin	竹林
Zhu Wen Gong Wen Ji	《朱文公文集》
Zhu Xi	朱熹
Zhu Zi Ji Cheng	《诸子集成》
Zhu Zi Yu Lei	《朱子语类》
Zhuang Zi (Chuang Tzu)	庄子
Zhuang Zi (Chuang Tzu)	《庄子》
Zhuang Zi Jie	《庄子解》
zi cai	自裁
zi de	自得
Zi Gong	子贡
zi jin	自尽
zi jing	自经
Zi Lu	子路
zi qiang	自戕
zi shi	自失
Zuo Zhuan	《左转》
Zuo Qiu Ming	左丘明

NOTES ON CONTRIBUTORS

Ronald A. Carson, Ph.D., is Director and Harris L. Kempner Professor at Institute for the Medical Humanities, the University of Texas Medical Branch, Galveston, USA.

Xunwu Chen, Ph.D., is Assistant Professor in the Division of English, Classics, Philosophy and Communication, University of Texas at San Antonio, San Antonio, USA.

Ruiping Fan, Ph.D., is Co-Managing Editor of *The Journal of Medicine and Philosophy*, Center for Medical Ethics and Health Policy, Baylor College of Medicine, Houston, USA.

Edwin Hui, Ph.D., is Associate Professor of Medical Ethics and Spiritual Theology and Associate Dean in charge of the Chinese Studies Program, Regent College, University of British Columbia, Vancouver, Canada.

George Khushf, Ph.D., is Assistant Professor in the Department of Philosophy, University of South Carolina, Columbia, USA.

Ping-Cheung Lo, Ph.D., is Associate Professor in the Department of Religion and Philosophy and Research Fellow of the Centre for Applied Ethics, Hong Kong Baptist University, Hong Kong.

Peimin Ni, Ph.D., is Associate Professor in the Department of Philosophy, Grand Valley State University, Allendale, USA.

Jing-Bao Nie, Ph.D., is Post-doctoral Fellow at the Center for Bioethics, University of Minnesota, Minneapolis, USA.

Qingjie Wang, Ph.D., is Assistant Professor in the Department of Philosophy at Oklahoma State University, Stillwater, USA.

Ellen Zhang, Ph.D., is Assistant Professor in the Department of Religion, Temple University, Philadelphia, USA.

Ruiping Fan (ed.), Confucian Bioethics, 299.

INDEX OF CHINESE TERMS

Ruiping Fan (ed.), Confucian Bioethics, 301–303.

INDEX

Ruiping Fan (ed.), Confucian Bioethics, 305–316.

Philosophy and Medicine

1. H. Tristram Engelhardt, Jr. and S.F. Spicker (eds.): *Evaluation and Explanation in the Biomedical Sciences.* 1975 ISBN 90-277-0553-4
2. S.F. Spicker and H. Tristram Engelhardt, Jr. (eds.): *Philosophical Dimensions of the Neuro-Medical Sciences.* 1976 ISBN 90-277-0672-7
3. S.F. Spicker and H. Tristram Engelhardt, Jr. (eds.): *Philosophical Medical Ethics.* Its Nature and Significance. 1977 ISBN 90-277-0772-3
4. H. Tristram Engelhardt, Jr. and S.F. Spicker (eds.): *Mental Health.* Philosophical Perspectives. 1978 ISBN 90-277-0828-2
5. B.A. Brody and H. Tristram Engelhardt, Jr. (eds.): *Mental Illness.* Law and Public Policy. 1980 ISBN 90-277-1057-0
6. H. Tristram Engelhardt, Jr., S.F. Spicker and B. Towers (eds.): *Clinical Judgment.* A Critical Appraisal. 1979 ISBN 90-277-0952-1
7. S.F. Spicker (ed.): *Organism, Medicine, and Metaphysics.* Essays in Honor of Hans Jonas on His 75th Birthday. 1978 ISBN 90-277-0823-1
8. E.E. Shelp (ed.): *Justice and Health Care.* 1981 ISBN 90-277-1207-7; Pb 90-277-1251-4
9. S.F. Spicker, J.M. Healey, Jr. and H. Tristram Engelhardt, Jr. (eds.): *The Law-Medicine Relation.* A Philosophical Exploration. 1981 ISBN 90-277-1217-4
10. W.B. Bondeson, H. Tristram Engelhardt, Jr., S.F. Spicker and J.M. White, Jr. (eds.): *New Knowledge in the Biomedical Sciences.* Some Moral Implications of Its Acquisition, Possession, and Use. 1982 ISBN 90-277-1319-7
11. E.E. Shelp (ed.): *Beneficence and Health Care.* 1982 ISBN 90-277-1377-4
12. G.J. Agich (ed.): *Responsibility in Health Care.* 1982 ISBN 90-277-1417-7
13. W.B. Bondeson, H. Tristram Engelhardt, Jr., S.F. Spicker and D.H. Winship: *Abortion and the Status of the Fetus.* 2nd printing, 1984 ISBN 90-277-1493-2
14. E.E. Shelp (ed.): *The Clinical Encounter.* The Moral Fabric of the Patient-Physician Relationship. 1983 ISBN 90-277-1593-9
15. L. Kopelman and J.C. Moskop (eds.): *Ethics and Mental Retardation.* 1984 ISBN 90-277-1630-7
16. L. Nordenfelt and B.I.B. Lindahl (eds.): *Health, Disease, and Causal Explanations in Medicine.* 1984 ISBN 90-277-1660-9
17. E.E. Shelp (ed.): *Virtue and Medicine.* Explorations in the Character of Medicine. 1985 ISBN 90-277-1808-3
18. P. Carrick: *Medical Ethics in Antiquity.* Philosophical Perspectives on Abortion and Euthanasia. 1985 ISBN 90-277-1825-3; Pb 90-277-1915-2
19. J.C. Moskop and L. Kopelman (eds.): *Ethics and Critical Care Medicine.* 1985 ISBN 90-277-1820-2
20. E.E. Shelp (ed.): *Theology and Bioethics.* Exploring the Foundations and Frontiers. 1985 ISBN 90-277-1857-1
21. G.J. Agich and C.E. Begley (eds.): *The Price of Health.* 1986 ISBN 90-277-2285-4
22. E.E. Shelp (ed.): *Sexuality and Medicine.* Vol. I: Conceptual Roots. 1987 ISBN 90-277-2290-0; Pb 90-277-2386-9

23. E.E. Shelp (ed.): *Sexuality and Medicine.* Vol. II: Ethical Viewpoints in Transition. 1987 ISBN 1-55608-013-1; Pb 1-55608-016-6
24. R.C. McMillan, H. Tristram Engelhardt, Jr., and S.F. Spicker (eds.): *Euthanasia and the Newborn.* Conflicts Regarding Saving Lives. 1987 ISBN 90-277-2299-4; Pb 1-55608-039-5
25. S.F. Spicker, S.R. Ingman and I.R. Lawson (eds.): *Ethical Dimensions of Geriatric Care.* Value Conflicts for the 21th Century. 1987 ISBN 1-55608-027-1
26. L. Nordenfelt: *On the Nature of Health.* An Action-Theoretic Approach. 2nd, rev. ed. 1995 ISBN 0-7923-3369-1; Pb 0-7923-3470-1
27. S.F. Spicker, W.B. Bondeson and H. Tristram Engelhardt, Jr. (eds.): *The Contraceptive Ethos.* Reproductive Rights and Responsibilities. 1987 ISBN 1-55608-035-2
28. S.F. Spicker, I. Alon, A. de Vries and H. Tristram Engelhardt, Jr. (eds.): *The Use of Human Beings in Research.* With Special Reference to Clinical Trials. 1988 ISBN 1-55608-043-3
29. N.M.P. King, L.R. Churchill and A.W. Cross (eds.): *The Physician as Captain of the Ship.* A Critical Reappraisal. 1988 ISBN 1-55608-044-1
30. H.-M. Sass and R.U. Massey (eds.): *Health Care Systems.* Moral Conflicts in European and American Public Policy. 1988 ISBN 1-55608-045-X
31. R.M. Zaner (ed.): *Death: Beyond Whole-Brain Criteria.* 1988 ISBN 1-55608-053-0
32. B.A. Brody (ed.): *Moral Theory and Moral Judgments in Medical Ethics.* 1988 ISBN 1-55608-060-3
33. L.M. Kopelman and J.C. Moskop (eds.): *Children and Health Care.* Moral and Social Issues. 1989 ISBN 1-55608-078-6
34. E.D. Pellegrino, J.P. Langan and J. Collins Harvey (eds.): *Catholic Perspectives on Medical Morals.* Foundational Issues. 1989 ISBN 1-55608-083-2
35. B.A. Brody (ed.): *Suicide and Euthanasia.* Historical and Contemporary Themes. 1989 ISBN 0-7923-0106-4
36. H.A.M.J. ten Have, G.K. Kimsma and S.F. Spicker (eds.): *The Growth of Medical Knowledge.* 1990 ISBN 0-7923-0736-4
37. I. Löwy (ed.): *The Polish School of Philosophy of Medicine.* From Tytus Chałubiński (1820–1889) to Ludwik Fleck (1896–1961). 1990 ISBN 0-7923-0958-8
38. T.J. Bole III and W.B. Bondeson: *Rights to Health Care.* 1991 ISBN 0-7923-1137-X
39. M.A.G. Cutter and E.E. Shelp (eds.): *Competency.* A Study of Informal Competency Determinations in Primary Care. 1991 ISBN 0-7923-1304-6
40. J.L. Peset and D. Gracia (eds.): *The Ethics of Diagnosis.* 1992 ISBN 0-7923-1544-8

41. K.W. Wildes, S.J., F. Abel, S.J. and J.C. Harvey (eds.): *Birth, Suffering, and Death*. Catholic Perspectives at the Edges of Life. 1992 [CSiB-1]
ISBN 0-7923-1547-2; Pb 0-7923-2545-1
42. S.K. Toombs: *The Meaning of Illness*. A Phenomenological Account of the Different Perspectives of Physician and Patient. 1992
ISBN 0-7923-1570-7; Pb 0-7923-2443-9
43. D. Leder (ed.): *The Body in Medical Thought and Practice*. 1992
ISBN 0-7923-1657-6
44. C. Delkeskamp-Hayes and M.A.G. Cutter (eds.): *Science, Technology, and the Art of Medicine*. European-American Dialogues. 1993 ISBN 0-7923-1869-2
45. R. Baker, D. Porter and R. Porter (eds.): *The Codification of Medical Morality*. Historical and Philosophical Studies of the Formalization of Western Medical Morality in the 18th and 19th Centuries, Volume One: Medical Ethics and Etiquette in the 18th Century. 1993 ISBN 0-7923-1921-4
46. K. Bayertz (ed.): *The Concept of Moral Consensus*. The Case of Technological Interventions in Human Reproduction. 1994 ISBN 0-7923-2615-6
47. L. Nordenfelt (ed.): *Concepts and Measurement of Quality of Life in Health Care*. 1994 [ESiP-1] ISBN 0-7923-2824-8
48. R. Baker and M.A. Strosberg (eds.) with the assistance of J. Bynum: *Legislating Medical Ethics*. A Study of the New York State Do-Not-Resuscitate Law. 1995 ISBN 0-7923-2995-3
49. R. Baker (ed.): *The Codification of Medical Morality*. Historical and Philosophical Studies of the Formalization of Western Morality in the 18th and 19th Centuries, Volume Two: Anglo-American Medical Ethics and Medical Jurisprudence in the 19th Century. 1995 ISBN 0-7923-3528-7; Pb 0-7923-3529-5
50. R.A. Carson and C.R. Burns (eds.): *Philosophy of Medicine and Bioethics*. A Twenty-Year Retrospective and Critical Appraisal. 1997
ISBN 0-7923-3545-7
51. K.W. Wildes, S.J. (ed.): *Critical Choices and Critical Care*. Catholic Perspectives on Allocating Resources in Intensive Care Medicine. 1995 [CSiB-2]
ISBN 0-7923-3382-9
52. K. Bayertz (ed.): *Sanctity of Life and Human Dignity*. 1996
ISBN 0-7923-3739-5
53. Kevin Wm. Wildes, S.J. (ed.): *Infertility: A Crossroad of Faith, Medicine, and Technology*. 1996 ISBN 0-7923-4061-2
54. Kazumasa Hoshino (ed.): *Japanese and Western Bioethics*. Studies in Moral Diversity. 1996 ISBN 0-7923-4112-0
55. E. Agius and S. Busuttil (eds.): *Germ-Line Intervention and our Responsibilities to Future Generations*. 1998 ISBN 0-7923-4828-1
56. L.B. McCullough: *John Gregory and the Invention of Professional Medical Ethics and the Professional Medical Ethics and the Profession of Medicine*. 1998 ISBN 0-7923-4917-2
57. L.B. McCullough: *John Gregory's Writing on Medical Ethics and Philosophy of Medicine*. 1998 [CiME-1] ISBN 0-7923-5000-6

Philosophy and Medicine

58. H.A.M.J. ten Have and H.-M. Sass (eds.): *Consensus Formation in Healthcare Ethics.* 1998 [ESiP-2] ISBN 0-7923-4944-X
59. H.A.M.J. ten Have and J.V.M. Welie (eds.): *Ownership of the Human Body.* Philosophical Considerations on the Use of the Human Body and its Parts in Healthcare. 1998 [ESiP-3] ISBN 0-7923-5150-9
60. M.J. Cherry (ed.): *Persons and Their Bodies.* Rights, Responsibilities, Relationships. 1999 ISBN 0-7923-5701-9
61. R. Fan (ed.): *Confucian Bioethics.* 1999 [APSiB-1] ISBN 0-7923-5723-X

KLUWER ACADEMIC PUBLISHERS – DORDRECHT / BOSTON / LONDON